FAWLEY'S
FRONT LINE

Fawley Refinery's Dennis F2/pyrene foam tanker LYM 259 (built in June 1952) on a training exercise involving storage tanks. (Dick Lindsay)

FAWLEY'S
A CENTURY OF FIREFIGHTING AND RESCUE
FRONT LINE

ROGER HANSFORD

The
History
Press

Dedicated to my father,
Ronald Joseph Hansford
1945–2012

Cover Picture Credits

Front: Fawley Volunteer Fire Brigade, *c.* 1939 (Keith Dyer); Army Fire Service Bedford TK/HCB Angus vehicle at Marchwood Defence Fire Station, *c.* 1970 (Trevor Fenn); a night-time training drill by the Defence Fire & Rescue Service (Trevor Fenn).

Back: Fawley Refinery's Dennis F2/pyrene foam tanker LYM 259 (Courtesy of Dick Lindsay).

First published 2014

The History Press
The Mill, Brimscombe Port
Stroud, Gloucestershire, GL5 2QG
www.thehistorypress.co.uk

© Roger Hansford, 2014

The right of Roger Hansford to be identified as the Author
of this work has been asserted in accordance with the
Copyright, Designs and Patents Act 1988.

British Library Cataloguing in Publication Data.
A catalogue record for this book is available from the British Library.

ISBN 978 0 7524 9857 7

Typesetting and origination by Thomas Bohm, User design
Printed in Great Britain

CONTENTS

LIST OF ABBREVIATIONS

AFFF	Aqueous Film-Forming Foam
AFS	Auxiliary Fire Service
AGWI	Atlantic, Gulf and West Indies Oil Company
ALP	Aerial Ladder Platform
APA	Auxiliary Plant Attendant
ARP	Air Raid Precautions
BA	Breathing Apparatus
BFSA	British Fire Services Association
BOC	British Oxygen Company
CAFS	Compressed-Air Foam System
CBRN	Chemical, Biological, Radiological, and Nuclear
CEGB	Central Electricity Generating Board
CFO	Chief Fire Officer
COMAH	Control of Major Accident Hazards
DFRMO	Defence Fire Risk Management Organisation
DF&RS	Defence Fire & Rescue Service
DIM	Detection, Identification, and Monitoring
gpm	gallons per minute
HART	Hazardous Area Response Team
HF&RS	Hampshire Fire & Rescue Service
HGV	Heavy Goods Vehicle
HMS	Her Majesty's Ship
HVP	High Volume Pumping Unit
ICU	Incident Command Unit
IFPA	Industrial Fire Protection Association

ISR	International Synthetic Rubber Company
LIFE	Local Intervention Fire Education
LPG	Liquefied Petroleum Gas
lpm	litres per minute
MBE	Member of the Most Excellent Order of the British Empire
MOD	Ministry of Defence
MRV	Multi-Role Vehicle
NFRDC	New Forest Rural District Council
NFS	National Fire Service
OBE	Officer of the Most Excellent Order of the British Empire
PC	Police Constable
PDA	Pre-Determined Attendance
PVC	Polyvinyl Chloride
RAF	Royal Air Force
RDC	Rural District Council
RTC	Road Traffic Collision
SEU	Special Equipment Unit
SFB	Southampton Fire Brigade
THA	Tactical Holding Area
VLCC	Very Large Crude Carrier

Note: The traditional term 'fireman' was superseded by 'firefighter' in 1992, and this is reflected in my text.

ACKNOWLEDGEMENTS

This book was made possible with help from serving and retired firefighters from the Waterside. Chas McGill at Hardley Fire Station allowed me access to his station and to a major emergency exercise, as well as giving me many contacts in the fire brigade. Trevor Fenn at Marchwood MOD Fire Station gave me a tour of his site, including a ride in the fire engine on 'blues and twos', and we spent several enjoyable hours sharing photographs and memories. Alan House, retired Deputy Chief Fire Officer for Hampshire, invited me to Fire Headquarters in Eastleigh and took the time to read my work, as well as offering invaluable research advice. I also met with Martin Rumsey, Malcolm Rumsey, Barry Browning, Derek Turner, Colin Partridge and Paul Freeman, and spoke on the telephone with Bill Farr. I would like to thank all the firefighters very much for their time and for the insight they gave me into their unique job.

Other local people have also been supportive. Mike Hocking at Geo Speciality Chemicals responded to my inquiry letter with useful information. At Waterside Heritage Centre in Hythe, Graham Parkes gave me considerable encouragement with my work and help with the early photographs, and Pam White provided interesting news clippings from her files. Jez Gale at the *Southern Daily Echo* assisted my archive work, finding me several decades' worth of clippings on Waterside fires. Janice Taylor at Herald Publishing promoted my project in her excellent local newspaper. Keith Dyer saw the *Waterside Herald* advertisement and got in touch to share his memories and an early photograph. Ernie Hartnell, formerly a local resident, responded to the advert from Australia! Pam Whittington wrote a 1998 book on the local fire brigade, and she discussed her research with me. I thank all of you, and hope this book gives something back to everybody in the area.

The work of several photographers enhances these pages: thanks to Steve Greenaway, Iain Kitchen, Matthew Leggott, Dick Lindsay, Roger Mardon, and Ken Reid (see Bibliography for individual websites). Clive Shearman and Malcolm Cheshire put me in touch with other photographers and sent me photographs from their collections. Hardley Fire Station and Hampshire Fire & Rescue Service kindly donated photographs, as did many individual firefighters.

Love and support from my mum and my wife Kaman also enabled the book to happen: thank you both!

INTRODUCTION

The rapid turnout of a fire appliance, with bright colours, flashing lights and loud sirens, is exciting to many people, not least when the humanitarian aspects of the vehicle's work are considered. The fire and rescue service has evolved over many years but the pride associated with the early origins of fire brigades has not diminished. The necessary qualities of discipline, caring and bravery remain unchanged to this day. The firefighter's role in the Fawley and Waterside area of Hampshire, one which presents firefighting risks of the highest level, is no exception. Firefighting tends to run in families, and many local firefighters were inspired to join by watching previous generations.

The preparations for civil defence during the Second World War marked a turning point in the organisation of fire brigades, but the arrival of a petrochemical industry brought the most significant change to provision on the Waterside. Today, fire and accident risks from industrial, commercial, domestic and rural causes are coexistent. The establishment of a new fire station at Fawley in 1977 provided immediate, round-the-clock cover for the area, but reduced opportunities and resources for existing fire stations. Some of the area's firefighters were employed full-time, and some were retained firefighters, responding from their homes or places of work, including in local industry. Fawley Fire Station, renamed Hardley in the 1990s, has worked hard to build good relations with local young people and to spread a fire-safety message.

A higher than average number of private fire brigades have operated on the Waterside, and the degree of collaboration between them, both in equipment and personnel, has gone largely unreported. The area is home to the UK's largest oil refinery, so a special focus on petrochemical firefighting is a particular feature of this book. Most of the industrial sites employ their own fire officers, drawing on additional workers at their plants for an emergency situation. The military presence at the Sea Mounting Centre (Military Port) in Marchwood involves the loading of ships with military

The 1990 water tender ladder G168 UPO was based at Hardley Fire Station from 1990 to 1997, and then at Lymington Fire Station. It later became part of the firefighting fleet at Fawley Refinery, and here carries the livery of Esso Fire & Response Group in place of HF&RS. (Iain Kitchen)

explosives, an operation covered by the Defence Fire & Rescue Service (DF&RS) and also featured in these pages.

This book is created to be of interest to firefighters, historians, Fire Service enthusiasts and local people alike. The incidents dealt with attest to the character of the locality and they tell a story of development and change. Fire brigade attendance at the same site over decades will find it in various stages of development, use or dereliction. I describe the way fire brigades dealt with major fires in Fawley Refinery – such as in 1985 and 2004 – and the way this site has prepared for the terrorist threat through large-scale exercises. My focus on vehicles offers a case study of the changing British fire appliance and shows how it has been shaped by commercial constraints. Against this trajectory I show that many fire vehicles on the Waterside differ from convention because of their adaptation to fulfil specialist roles.

My own great-grandfather served at the Auxiliary Fire Station based behind the Falcon Inn in Fawley village, and he probably triggered my interest. As a young child growing up in the area, I saw many fire engines and wondered where they were going and whether there was more I could know about them. At an Esso open day in 1989, I was one of the children encouraged to compare a local authority with a Refinery fire engine. I continued to encounter the fire brigade as part of daily life, going to the Traveller's Rest public house in the mid-1990s only to find it on fire: we stayed to watch as two Hampshire machines roared up and their crews tackled a blaze in the kitchen! During the preparation of this book in September 2011, I left home early one Sunday morning and found myself following Hardley's water tender along the A326 as the crew responded to a fire at Marchwood Scientific Services.

I would like to extend particular thanks to Chas McGill and Trevor Fenn for their support of this project. Through them I was able to meet whole-time and retained firefighters, both serving and retired. I am indebted to Alan House who improved my work and whose own publications provided a broad background to the topic. The *Southern Daily Echo* archive was very useful in showing me how incidents were reported to local residents, and the study of this newspaper itself forms a historical sweep. The highlight of my research was being able to observe Exercise Shannon first-hand, and I extend my appreciation to all the personnel involved at the petrochemical complex on 5 May 2012. In publishing this book, I hope the work of the Waterside fire brigades will be better appreciated and that respect for them will continue to grow as they manage risk in the service of their community.

1

CADLAND ESTATE AND THE EARLY FIRE BRIGADES

At the start of the twentieth century, much of the parish of Fawley was rural in character, and the area consisted of villages, farms, lanes, farmland and woodlands. Significant buildings were the Norman church at Fawley and the Tudor stone castle on Calshot Spit. By this time the Drummond family was well established at the Manor of Cadland, having extended the house originally designed by Henry Holland and set in 'Capability' Brown parkland. The records of a fire at the estate farm in the late nineteenth century give a glimpse of the practice of firefighting in the parish at this time. My research on the fire is sourced from two accounts of the event, which I refer to as 'Report 1' and 'Report 2', both included as cuttings in the Drummond family's scrapbook dated 1894. It has been difficult to date the incident precisely, but it could have been the fire at 'Cadlands Farm, Cadlands' attended by Southampton Fire Brigade on Sunday 1 March 1885, as documented in their records (SC/F 1/1).

Report 1 describes a 'destructive fire at Cadlands Home Farm'. This was 'of a very extensive character' and destroyed five ricks of hay, six of barley, four of wheat, one of barley straw and one of ferns, later spreading to the granary and cart house, which almost completely burnt down. The initial alarm was raised at 8 p.m. on the Saturday by a Mrs Sarah Fry, steward to Edgar Atheling Drummond and servant to the farmer, Mr Hogg. At this, 'the farm servants and others quickly mustered, and there being a plentiful water supply, the fire was soon extinguished'. Although a watch was kept on the yard, fire broke out again at 1.30 p.m. on the Sunday. This was attributed to brisk winds fanning embers from the previous day's fire, which had been spread out in the open close to other ricks. Report 2 stated that in total some eighteen ricks were on fire within ten minutes of the alarm being raised in the village by one of the farm boys.

Report 1 details which fire brigades responded, and how they were called. Two engines were sent from nearby estates, one by Count Batthyhany of Eaglehurst Castle and another by Lord Henry Scott of Beaulieu. The naval vessel HMS *Zealous* was anchored off Netley, and dispatched two engines 'manned by marines and sailors, and taken in boats up Cadlands Creek, to the scene of the fire'. These engines, when on land, were probably drawn by horses, with pumps operated manually or using steam. Report 1 suggests the engines were mobilised when the fire was sighted: 'The conflagration was of course very great, lighting up the whole country round.' More sophisticated methods of despatch were also in use, however, as this report informs us that a telegram was sent to Superintendent Gardner of Southampton Fire Brigade (SFB). Despite a slight difference in spelling, there was a Mr W.H. Gardiner in charge of SFB from 1876 until the mid-1880s, the term 'Superintendent' being used instead of Chief Fire Officer at that time. The officer and his personnel took the shortest route to the fire by catching the Hythe Ferry across Southampton Water. Meanwhile, their engine was sent around by road through Redbridge and Totton, but returned to base after meeting Mr Hogg's messenger in Totton, who stated the fire was too seriously advanced for the engine to be of use. This left the Southampton firemen to assist the crew from Beaulieu, whose engine was the first to arrive. Report 2 said it was 'thought necessary to telegraph for the Sappers and Miners', and that the Sappers had arrived at Hythe before receipt of a second telegraph saying they were no longer needed. This may refer to the Southampton engine described in Report 1.

Report 1 gives an idea of the firefighting methods employed at the scene. The fire had spread too rapidly for those present to remove the carts from the cart house before it was engulfed. However, they did remove cattle from the cattle house and sprayed water on the farmhouse and other buildings to prevent them from igniting. Several water sources were available, all of which were utilised. There was a large tank of water 300–400m from the rick yard, and Report 1 states: 'Two lines [of people] were formed from the yard to the tank, the one side passing empty buckets down from hand to hand, and returning full buckets of water in the same way on the other sides, and thus some of the engines were kept supplied with water.'

In this task, the Southampton and Beaulieu teams 'did exceedingly good work, the men behaving themselves admirably', and they were supported by many volunteers. Among the volunteers were important local figures including Mr Hogg, Mr Drummond, Mr Jenkinson, Revd Unwin, Dr Stephenson and Mr Perkins, and the same report states, 'other gentlemen came, each lending a hand at the pumps or hose'. Meanwhile, Count Batthyhany directed the sailors and others in showering water onto the burning ricks from the farm pond, where one of the *Zealous* engines was working. Twenty men from HMS *Zealous* worked through the night with water from a tank in the roof of the corn mill, managing to save the corn and a threshing machine. The first team worked until midnight, when they were relieved by several other detachments from the ship, each led by an officer. This coheres with the Southampton Fire Brigade record, which mentions the attendance of private and manual engines not of SFB, and says the Cadland blaze was extinguished by 'Firemen and Strangers'.

Despite the efforts of all involved, the impact of the fire was huge, doing damage estimated to cost £15,000, and ruining a good harvest along with the farm buildings.

A tragic fire at Cadland Farm on 11 May 1923 was triggered by a boiler explosion and caused £5,000 worth of damage. Southampton Fire Brigade's motor pump No. 3 attended the call, which had been raised by Leycester Meyer of Fawley. Alfred James Eldridge, aged 29, was killed in the incident along with his horse and three calves. (Waterside Heritage)

Smoke continued to issue from a hayrick at the scene on Monday, leading local people to believe the fire was still burning then, and there were unfounded rumours in Southampton and Beaulieu that Cadland House itself had been on fire, and even that fires had occurred simultaneously at Broadlands and Netley.

Enquiries were made into the origin of the blaze, and the police officers Superintendent Troke and Sergeant Fox concluded that the cause of the initial fire on the Saturday was accidental. Report 2 suggested spontaneous combustion as a possible cause of the fire on Saturday, and questioned whether 'the crime of incendiarism' had played a part in the fire's rapid spread on Sunday. This was thought unlikely given that watchmen had been on guard all morning, and also that Mr Drummond was well-respected in the area. He was not a 'hard landlord' but 'a most affable and kind gentleman, taking delight in and doing his utmost to promote everything that is for the improvement of the estate'. The author was 'not afraid to assert' that he was 'more regarded and respected by all classes' than any other landlord in the country, and could therefore 'scout the idea with scorn that anyone of the parish of Fawley could have been guilty of such a dastardly outrage'. A final suggestion was that a 'fanatic' of the 'destestable Trades' Unions' may have started the fire, but no such guilty party could be found. Put in context of the scientific approaches to fire investigation used in modern times, it is amusing that the accidental verdict was based largely on the character of the property owner. I suggest the failure to keep burnt embers away from fresh crops on the Saturday, and the rejection of the Southampton engine on the Sunday, leaving its crew unequipped and depriving the incident of an extra pump, were significant errors. Interestingly, the nineteenth-century reports do not question the allegiances or competence of those left to watch the yard before the main outbreak on Sunday morning!

The accounts of the Cadland Farm fire in Reports 1 and 2 are significant because they raise issues that have been important for firefighting in the parish – and in this location – over the last century. The reports evoke pictures of a rural landscape where farming was paramount, and rural firefighting has been commonplace in the surrounding area ever since. After development as an industrial site, the land at Cadland often saw fires and incidents where different fire brigades would respond and work together. The military are mentioned in the Drummond reports and they were involved with firefighting and rescue in the wider vicinity at different times; for example Southampton Fire Brigade recorded assistance from soldiers and sailors at a fire in Bourne Hill Cottage, Fawley, on 10 September 1916. In the 1880s many people from the community were prepared to help at Cadland on a voluntary basis, showing a willingness to serve in the face of danger which has not disappeared in the present day. Some of the firefighting techniques used in the farmyard, such as passing buckets down a line of people, differed little from those used since ancient times and employed, for example, at the Great Fire of London in 1666. Such techniques continued to be used into the early 1900s. The methods used to prevent fire spreading anticipated modern techniques, but the inefficiency of communication between fire teams contrasts strongly with the present. Finally, although four engines were called to a fire within the parish, all of them came from stations outside the boundary and none of them were from Fawley.

2

MODERNISATION AND WAR: THE BEGINNINGS OF ORGANISED FIREFIGHTING IN FAWLEY

The first significant signs of change in Fawley occurred in 1920 when the Atlantic, Gulf and West Indies Oil Company (AGWI) started building a small refinery for bitumen and bunker fuels on land from the Cadland Estate. On 19 July 1921, SFB recorded attending a fire, with Engine No. 3, which had broken out in 500 out of the 1,200 sleepers piled beside the Totton railway line that were 'to be used for the new Fawley line' (SC/F1/1). This call-out heralded the arrival of the railway on the Waterside, meaning the area would no longer be isolated from development.

A fire brigade began in Fawley's neighbouring parish of Hythe in 1919, moving to a purpose-built fire station in New Road in 1927. Pam Whittington's *Hythe Fire Brigade: A Local History* (1998) tells the story of this brigade, including incidents they responded to within the Fawley area. In contrast, it would be another half-century before the parish of Fawley had its own purpose-built and permanent local authority fire station. Yet the area had a successful volunteer fire brigade before the legal requirement to provide a Fire Service came in the late 1930s. This chapter covers developments underlining Fawley's need for a local authority fire station and describes the measures taken in the meantime to save life and property from fire. From the 1920s to the 1960s the degree of petrochemical risk was to increase unrecognisably and special arrangements were needed for the threat of armed assault and invasion by Nazi Germany. However, it was not until the 1970s that the new fire station was built.

The Fawley Volunteer Fire Brigade in the 1930s

During the first decades of the twentieth century there was no fire brigade in the area. No legislation required such provision and no piped water supply was available.

The population relied on pumps and wells for domestic purposes and the authorities jostled to avoid responsibility. The minutes of the Fawley Parish Council, held at the Hampshire Record Office in Winchester (Shelfmark 25M60/PX1), show how the need to improve firefighting capacity was recognised in the parish as early as 1927, but it was initially difficult to obtain necessary backing from the New Forest Rural District Council (RDC). The entry for a meeting held on 21 March 1927, at 7 p.m. in the Public Hall, reads:

Fire Hydrants

Mr. G. Musselwhite brought forward the question of Fire Hydrants, stating that in view of the recent disastrous fire at Copthorne, and difficulty of obtaining supplies of water in case of fire, he considered that it was advisable that Fire Hydrants should be fixed in the village. Mr. Maclean explained what had been done by the Agwi Co. regarding Hydrants for their works and a letter was read from the clerk to R. D. C. explaining necessary steps to be taken to obtain Hydrants, this matter was left to the Parish Council.

It is interesting that firefighting practice in AGWI was seen as a benchmark for local improvements. In 1931, SFB attended fires at Ashlett Cottage, Fawley, and at Hubert Scott-Paine's British Powerboat Company in Hythe, involving twelve powerboats. The Fawley minutes for 1931–32 recorded the Parish Council 'pressing' the RDC to 'act' and to help fund a water supply to Blackfield, Langley and Spratts Down. Clearly firefighting would have been just one of many uses for piped water!

By 27 March 1933 the situation had improved and Fawley Parish Council was 'pleased to report the following progress on parish affairs during the past year'. Regarding the water supply to Rollestone, Blackfield, Langley and West Common, arrangements were 'well under way', and with help from the RDC and Southampton Borough Corporation this had involved minimal expense for the parish. However, later entries show that local people were sometimes reluctant to take advantage of the water supply, particularly at Langley, West Common and Ashlett Creek. The 1933 report stated, 'application will also be made to include hydrants'. The second point of the report merits full quotation as, with acknowledgement of financial constraints, it shows how seeds were sown for the first organised fire team in Fawley:

Fire Protection

The rapid development of building in the parish makes the provision of adequate fire protection facilities necessary.

   Within its limited means and with a desire to restrict expenditure the Council cannot do much towards providing all the appliances desirable but the formation of a volunteer fire brigade has been under consideration and members of the public who are interested in this matter should give their names to Mr. Dobson the chairman of the Fire Appliances Sub Committee.

The Parish Council's concern for firefighting provision was prescient given the major incident at the AGWI Refinery on 12 March 1935. Headlines from the *Southern Daily Echo*'s report of

Fire at Copthorne House, 5 November 1926. Furniture has been removed to safety on the lawn. An oil lamp started the fire, which severely damaged the dwelling owned by AGWI Corporation and accommodating Superintendent Mr Demoulins and Assistant Superintendent Mr Wishart. (Waterside Heritage)

Firemen at the Copthorne House fire using a soda–acid operated first-aid hose attached to the fire appliance, as well as foam fire extinguishers. The extinguishers would have been ineffective for a house fire! In the left foreground is Captain W. Burt, Chief Officer of Hythe Fire Brigade. Firemen from Hythe, Brockenhurst and Exbury attended this fire, and the vehicle shown is probably the Exbury one. (Waterside Heritage)

Petrol tank fire at the AGWI Refinery, 12 March 1935. (Waterside Heritage)

the next day called the fire 'alarming', with homes 'rocked' by a violent explosion, a petrol tank 'turned into a roaring, fiery cauldron', one man making a 'miraculous escape', and the firemen engaged in 'twelve hours' battle'. The tone of the article is amusing to modern eyes:

> Fawley's night of fear and tension is over, but the memory of those anxious hours is one that will ever remain with those who went through agonies of suspense. Through long, sleepless hours women watched and waited while their menfolk fought a terrifying fire … From the windows and doors of their homes clustered about the vast plant they were silent witnesses of the most alarming waterside fire in the history of the port … the womenfolk shuddered when the alarm sirens screeched at 5.35 last evening, just when they were expecting their bread-winners home for their evening meal.

Aside from gender inequalities, the reporters were right to be concerned; this type of incident was a first for the area. Sixty firemen were involved, including AGWI teams, with seven of them wearing asbestos suits and gas masks. Twenty-two water jets were used, two on the burning tank and the rest on nearby tanks put in danger by changes in wind direction. Some 20 tons of foam powder and 8,000ft of hose were used. Southampton Fire Brigade, led by Chief Officer Hayward, took twenty-five minutes to get there to support crews from Hythe and Brockenhurst, and airmen from the RAF base at Calshot were put on stand-by to attend. The workers Albert Andrews and John Wheeler were injured in the initial explosion, but Alan Read escaped unharmed. The *Echo* described the AGWI petrochemical installation as the largest in England, worth £2 million, with 112 tanks holding 227 million litres of petroleum products. Pleasingly, the fire did not spread further, and no lives were lost.

A year later, the Parish Council minutes again raised the question of fire protection, this time suggesting a Fawley Fire Brigade be formed. Quotations testify to the success of this volunteer organisation, which operated 1936–39:

Aftermath of the fire at the AGWI Refinery in March 1935, showing the remains of the devastated tank. (Waterside Heritage)

30th March 1936
Formation of Local Fire Brigade

Mr H. H. Edwards spoke of the suggested formation of a Fawley Fire Brigade for which volunteers were required.

He states that the Agwi Petroleum Corporation Ltd. has put their Leyland Fire Engine at the disposal of the Council for fighting local fires and that it was desired to have trained men in different parts of the parish to meet any future contingencies.

Water to Spratts Down

Mr G.R. Allan dealt with the necessity of water at Spratts Down both for domestic purposes and in case of emergency for firefighting.

At the Parish Council meeting on 21 March 1938, at 7 p.m., it was reported that the water supply to Spratts Down had reached a 'satisfactory conclusion', with two extensions being recommended for Langley. The following statement pertained to the Fire Brigade:

With the continued growth of the Parish, the subject of maintaining a satisfactory Fire Brigade becomes more important. Now that we have been successful in obtaining volunteers who have undergone training and qualified to take an engine to any fire, it has become necessary to provide suitable uniforms and equipment, which, has accordingly been done.

Clearly this brigade was adequately staffed and well-organised, with the local petrochemical industry providing vehicular support. It is tempting to conclude but is not certain that the 1935 tank explosion had precipitated their formation, demonstrating a degree of symbiosis between the petrochemical plant and local firefighting provision.

Firefighting Arrangements During the Second World War, 1939–45

Evidence of the growing threat of war can be seen from the Parish Council minutes of the late 1930s. This threat would lead to changes in the nationwide organisation of firefighting, meaning that several different fire brigades would operate in Fawley over just a few years. The 1938 report said a special Parish Council meeting had been held with regard to Air Raid Precautions (ARP), and that 'arrangements have been made for the formation of the different sections and suitable personnel have been recruited', but that 'both men and women' required further training. Initially, the responsibility for firefighting passed to the RDC. On 30 March 1939 at 7.30 p.m. in Jubilee Hall, 'A vote of thanks in appreciation of their services was expressed to the members of the Fire Brigade which has now been taken over by the higher authority'. The *Annual Report for the Year* read:

Fire Brigade

In accordance with an Act of Parliament of last year, The Fawley Fire Brigade has been taken over by the Rural District Council, and will, in future, be administered by that body.

Credit has been expressed for the work done by the Brigade and also with regard to their keenness and ability.

It is pleasing to see that, according to these minutes, Fawley's first fire team were strong and well regarded in their community. They had already dealt with some challenging incidents, not least the severe fire at Cadland House in 1937, at which furniture was removed onto the lawn for salvage purposes. Fawley was not alone in tackling this fire, as crews attended from other stations, including Brockenhurst. The organisational situation at Fawley had moved on by the Parish Council meeting held on 18 March 1940 at 7.45 p.m. in Jubilee Hall. Signed by the chairman, Mr Cave, the Minutes stated:

The transference of the Fawley Fire Brigade to the Local Authority, namely, the New Forest Rural District Council, has been accomplished and your Council has been recompensed for all expenditure incurred since its formation.

Firemen spraying water at the exterior of Cadland House at a fire in 1937. Both brass and leather helmets are being worn, suggesting that two different fire brigades were working together. (Waterside Heritage)

Cadland House fire, 5 May 1937. As at Copthorne, furniture has been removed onto the lawn, and water is being sprayed at the building. Southampton Fire Brigade estimated the cost of damage to the West Wing at £23,400. (Waterside Heritage)

Our help is now required in providing more personnel for the A. F. S., (Auxiliary Fire Service) and any men or women who can find the time for this work of national importance should get in touch with the Firemaster, Mr H. H. Edwards at Blackfield.

Review of Past Year's Work

Stirrup pumps are now being provided and the training of crews will be completed as soon as sufficient volunteers come forward.

At the Parish Council meeting on 31 March 1941 the discussion covered householders' co-operation with tasks including firefighting, first aid, waste collection and food production, and 'practical demonstrations were given in the use of stirrup pumps and life saving'. Once again the local community was involved in a national effort which the Council 'earnestly' hoped would attain 'victory' and a return to 'the happier ways of peace' before the next annual meeting.

In fact, war was to continue for another four years. One teaser from the 1939 minutes is that 'prior to handing over to the District Council, a site had been chosen for a fire station which would also serve for A.R.P. purposes'. Disappointingly the minutes do not state where this was, but it is known in the area that a temporary fire station existed behind the Falcon Inn at Fawley for AFS use. Photographs show the building in dereliction, with a sign still showing at the door to address the firemen: 'All officers and men in charge of crews report here.'

My own great-grandfather, Charlie Hansford, served from this station, and it seems he had responded to Mr Cave's 1940 request for personnel. At the family home built by Charlie we still have his belt, axe, and helmet. We also have a stirrup pump thought to be of ARP issue;

This building behind Fawley's Falcon Inn was used as an AFS Fire Station during the Second World War. (Waterside Heritage)

A notice on the fire station behind Fawley's Falcon Inn left over from the Second World War, showing the station was used as a reporting point. (Waterside Heritage)

Charles Hansford, who served as a fireman in Fawley during the Second World War, pictured in uniform outside the house he built in 1923. His leather helmet suggests the photograph was taken between March 1939 and March 1940. (Roger Hansford)

these were made for 'Fire Watcher' or 'Fire Guard' duties in a particular street or location, and Alan House believes Charlie may have obtained his pump from someone else. It has passed down into family lore that when ever Charlie cycled from Hardley to the Falcon station the Home Guard would be blocking the road with a tree and shouting 'who goes there', which he thought unnecessary and ineffective. He remembered leaving the site of a severe pine tree fire caused by incendiary bombs. Charlie also attended a boathouse fire at Hythe where the firemen were jeered at by civilians; the latter were quick to return to the air-raid shelter when the sirens re-sounded, but the firemen worked on through the raid.

Other local people have memories of the wartime firefighting arrangements. Ernie Hartnell's father was a member of the AGWI Fire Brigade, which expanded in January 1940 to provide twenty-four-hour cover. When Ernie was a toddler his mother would take him in his pushchair from their Fawley home to the refinery to deliver his dad's lunch. All the firemen would gather around and toss a coin to see who got the job of helping Mrs Hartnell push Ernie back up the hill to the village! Ernie has lived in Australia since the 1960s. Keith Dyer's father and uncle both served in the Fawley volunteer fire team. At the time of the photograph below, Keith was aged 3 or 4, and he remembers going to a Christmas party with the team and seeing Mickey Mouse films there for the first time; also using his father's axe for chopping wood in later years. It is tempting to speculate that the solid-tyred fire tender shown in the photograph is the one lent by AGWI to the Fawley volunteers. Certainly from the parish minutes there was much overlap between petrochemical and local authority firefighting arrangements, and this has continued to the present day.

Fawley Volunteer Fire Brigade in the AGWI Refinery, *c.* 1949. Keith Dyer's father, Jack, is third from left, and his Uncle Frank is third from right (back row). The suited gentleman in the centre of the photo, possibly Chief Fire Officer Heptinstall of the newly-formed NFRDC Fire Brigade, is flanked by senior fire officers whose rank is denoted by the braiding on both shoulders; their two deputies standing behind have braiding on their left shoulders only. The crew's badges have been purchased from the National Fire Brigades' Association, which saved the Fawley Brigade from producing their own. (Keith Dyer)

The Fawley Fire Brigade was established with crews, accommodation, and the use of a vehicle, but at that point their management was transferred to another authority. The legislation alluded to by Fawley Parish Council in their Minutes was the Air Raid Precautions Act 1937, creating the AFS, and the Fire Brigades Act 1938. The latter created the New Forest Rural District Council Fire Brigade, formed on 29 January 1939. The *c.* 1939 photograph of the Fawley team shows their brigade on the cusp of absorption. As Pam Whittington (1998) explains, a survey by Chief Officer Hayward of Southampton Fire Brigade valued Fawley's Brigade at £132 6s 0d, which was the least of all the New Forest stations to be absorbed into NFRDC. After the organisational change, Fawley worked more closely with Hythe, training with them and standing by there sometimes. Alan House (2001) reports that by 1 April 1939, twelve volunteer firemen had transferred to New Forest from the Fawley Parish Brigade. By November of that year, Fawley had a first-line (full-time) crew of six plus a watch room attendant, and their equipment comprised one light trailer pump, one tender for towing and one mobile water carrier in the form of a lorry with a water dam. By 1941 the station had two car-towing vehicles: a 1935 Sunbeam Saloon registered AAA 282 pulling an Apex large trailer pump and a Morris Tourer registered OU 5234 pulling a Beresford light trailer pump. Initially, equipment was kept in the refinery but used outside by volunteers, until accommodation could be found in the local community. Captain Edwards was asked to make available his own car, an Armstrong Siddeley, along with garage space, and he later became a despatch rider. He had stayed on as officer in charge since the formation of the Parish Brigade.

Alan House's *Forest Firemen* (2001) tells the story of the New Forest RDC Fire Brigade, and documents the incidents they attended, including heathland and domestic fires, and their response to air raids in the locality. Of particular note was a serious incident at Cadland Estate on 6 September 1940. Firemen from AGWI also attended this large fire in undergrowth close to their plant, which started during the early hours of the morning. Afterwards the New Forest Chief Fire Officer, Mr Turner, was asked to resign for mismanaging the despatch of reinforcement crews, and he was replaced early in 1941. House's research suggests the maintenance of the Fawley section, in terms of crew numbers and standards, communication, equipment and accommodation, was more problematic than at many other stations in the New Forest. This finding contrasts with the optimistic discussion of the local Fire Service in earlier years found in the Parish Council minutes, but it chimes with a dip in firemen's morale seen nationally at this point of the war. This is despite the fact that NFRDC paid its firemen whilst Fawley parish did not.

The National Fire Service (NFS) was formed on 18 August 1941, the south of England being covered by Region No. 6, which answered ultimately to the Home Office. Fawley was designated as NFS station '3W' within B Division of the No. 16 Fire Force, giving it the identification number 16B 3W. There was also an NFS station based at the Flying Boat pub in Calshot, which operated until 18 April 1945. Alan House told me that in addition to fire stations there were two action stations in the area, one at Marchwood staffed by Totton's crews and one at Blackfield staffed from Hythe or Fawley. During air-raid alerts, fire crews responded from the parent station to the action stations, which stored additional equipment to use for bomb attacks or fires in the vicinity. Further information on the organisation and employment of the Fire Service at this time is available in publications by Alan House, John Leete, and by the Fire Brigade Society (see Bibliography). Suffice to say that the provision of a National Fire Service in Britain was unprecedented in 1941 and has not been seen since. This attests to the high risk presented to the civilian population by the war with Germany and her allies. In the Fawley area, the AGWI Refinery was particularly under threat during this time, and Pam Whittington (1998) notes that it was bombed several times in 1940–41, and that fires were started deliberately in the New Forest to provide a smoke screen for the site, or to act as decoy targets for German bombers. The Norman church building at Fawley was bombed in 1940, as was the South Western Tar Distillery in Totton in 1941.

Post-war Fire Cover on the Waterside

Once the Second World War was over, the government decided to disband the NFS and return the Fire Service to local authority control, but at a higher level than the pre-war district brigades. Hythe's current fire station at New Road was opened in 1947 and, as a result of the Fire Services Act of that year, became absorbed into the Hampshire Fire Service, which came into being on 1 April 1948. Along with Beaulieu, Hythe was the main fire station covering the Waterside – the area between the New Forest and Southampton Water, with Marchwood in the north and Calshot and Lepe in the south. The wartime Fawley station was closed and its equipment removed on the demise of the NFS.

The military maintained a post-war presence in the area, and this in itself presented fire risks. A base near Hythe for minesweepers and torpedo boats, known as HMS *Diligence*, opened in 1953 and suffered a fire on 17 August that year. In 1955 the site underwent an inspection looking at the fire risks it posed. More than fifty craft at any one time, each valued at £250,000, were to be moored in Southampton Water, opposite the naval base, and some would be brought ashore and waterproofed using liquid with a flashpoint of 15° F. In addition, the pier at HMS *Diligence* could withstand weights of only 4 tons, which was too little to support a fire appliance.

A. W. Paramor, Hampshire's first Chief Fire Officer, wrote a letter on 12 July 1955 stating that the county could supply lightweight pumps and hoses for ship firefighting – so long as the Admiralty could provide a craft to ferry these to any burning vessel offshore from HMS *Diligence*. Captain Stanning, secretary to the Commander-in-Chief at Portsmouth Dockyard, replied on 29 September that as no such craft was available a fireboat would be needed. Although a fireboat was kept by the fire brigade in Southampton, they were a separate brigade under the 1947 legislation. Their vessel, named *Fireboat 27*, was a Second World War estuarial fireboat first issued to the NFS and in service with SFB from 1948 to 1963. Hampshire Fire Service could request the fireboat but could not include it within a Pre-Determined Attendance (PDA) unless the Admiralty took this up directly with Southampton Fire Authority. Chief Officer Hayward of Southampton was prepared to supply the fireboat but on a chargeable basis.

The Hampshire authority rejected Hayward's offer in December, pointing out that vessels equipped for firefighting could be sent from the RAF base at Calshot, or from Marchwood Military Port when crews were there training. The Calshot vessel was a continuously crewed air-sea rescue launch with firefighting equipment, and a second launch was also available from there. The launches could be despatched more quickly than the Southampton fire-boat, which was sometimes away from its moorings to cover flying boat take-offs. There were two commercial tugs at Marchwood fitted with firefighting equipment, one diesel and one steam, and other craft able to carry fire pumps. All the sites along Southampton Water, including Marchwood Power Station which was then under construction, could be accessed by road using Hampshire's own appliances. By January 1956, Mr Paramor had been replaced by Mr Bowles as Hampshire's Chief Fire Officer (CFO). Captain Stanning still 'intended to press for the services of the Southampton Fire Boat'. However, on 30 January, Chief Fire Officer Bowles wrote the following memo, marked 'personal':

> HMS *Diligence* is a special risk for which the County Council are under no obligation to make provision, and it is therefore for the Admiralty to take such steps as they con-sider appropriate, and we should be careful not to appear to tell them how to face up to their own responsibilities … I suspect that Southampton Corporation may be using the Admiralty as a means to compel us to contribute towards the cost of their fireboat …

The sense of animosity between local authority brigades that would later amalgamate is intriguing. The documents reveal the firefighting resources available on the Waterside during the 1950s. The RAF at Calshot had been called on several times before for major incidents in the area, and they provided Hampshire with a good argument in this case. There was an air base at Calshot from March 1913 to April 1961. An AFS exercise code-named 'Ruffit' culminated at HMS *Diligence* on 21 September 1958, addressing the issue of offshore firefighting.

The mention of flying boats in the area is interesting, as the Imperial Airways flying boat *Connemara* had caught fire off Hythe just before the Second World War whilst refuelling from the launch *Mexshell*, killing a launch hand called H. Vincent. Pam Whittington (1998) provides an account of this incident, and she also mentions that the commercial fire brigades operating in the area had included those of Imperial Airways, British Overseas Airways Corporation, British Powerboat Company, Supermarine and Burt, Boulton & Hayward of Totton. Whittington also mentions the 1958 Solfire plan which involved the Hythe and Beaulieu Fire Stations in the planned response to a ship emergency in the Solent. By the mid-twentieth century, fire cover on the Waterside had been provided by a mixture of mili-tary and commercial brigades working alongside those of the local authority.

The Growth of the Petrochemical Industries from 1950

The construction of a vast new Refinery began on the site of Cadland House in 1949, and took more than two years to complete. The Anglo-American Oil Company had engulfed the AGWI company and decided to site its main refinery on a huge tract of the Drummonds' land, renaming itself the Esso Petroleum Company Ltd when the refinery opened in 1951. The main building contractor, Foster Wheeler Ltd, had its own fire brigade, which became the Esso Fire Brigade when the refinery went on-stream. Technologies employed for the refining of oil have evolved over the years, but the principles have remained the same, as has the scale of operations. The entire site covers 3,200 acres of land. In 2012, the company website said the Fawley Refinery was the largest in the UK and one of the most complex in Europe, with a mile-long marine terminal handling 2,000 ship movements and 22 million tonnes of crude oil and its products annually. The site at that time was processing 330,000 barrels of crude oil per day and could provide up to 20 per cent of UK refining capacity.

The Control of Major Accident Hazards (COMAH) legislation regulates operations at the Fawley Refinery. It is important to understand the basic refining processes in order to appreciate the possible firefighting risks. On arrival at the refinery, the crude oil is distilled so as to be split into the useful components of gas, petrol, aviation fuel, diesel and lubricating oils. The lower distillates of fuel oil and bitumen are no longer produced at Fawley. Primary distillation requires the crude oil to be heated to 580°C, whilst the site's catalytic cracking unit uses temperatures of around 725°C to break heavier fuels down into lighter ones to meet consumer demand. Esso Chemicals began operation next to the refinery in 1966, later renamed Exxon Chemicals, and a merger formed the Exxon-Mobil Corporation in 1999. Fed by naphtha from the refinery, the chemical plant produces

Inside Fawley Refinery, a test of one of the early portable monitors such as were later carried on the Bedford TK tail-lift vehicle (see Chapter 5). The monitors were designed and made in the refinery before they were available commercially from manufacturers such as Chubb and Strebor. (Barry Browning)

850,000 tonnes of products every year. One key substance is methyl ethyl ketone, a solvent in paints and adhesives. Another is halobutyl rubber, used widely in the production of tyre linings. Whilst the former requires a catalyst of sulphuric acid, the latter requires chlorine or bromine, both of which are hazardous products that must be transported to the site. In turn, sulphur is recovered in the refining process and this is carried away by road as a molten liquid.

A fuel tanker loading facility known as Hythe Terminal opened in 1960. The refinery supplies Esso road tanker terminals nationwide via ship and pipeline, in addition pumping aviation fuel to Heathrow Airport. Some distribution takes place using the railway, which also served as a passenger line during the early 1960s. Between Hythe Terminal and Southampton Water a number of additional premises have been built which deal with the production and disposal of chemicals, or the distribution of gases. The British Oxygen Company (BOC) extracts oxygen from the atmosphere to supply the refinery with compressed air; it also transports argon and nitrogen to customers in liquid form using chilled tankers. Calor Gas and Flogas transport Liquefied Petroleum Gas (LPG) by bottle and by road tanker from their depots in Cadland Road. The labelling of road tankers has been addressed by legislation including the 1978 Hazardous Substances Regulations and, more recently, the 2009 Carriage of Dangerous Goods Regulations. This means the vehicles are required to display on a prominent orange panel the United Nations product number for the load carried together with an advisory code for the fire brigade.

A decontamination area set up during a simulated chemical incident. This was one of the displays for the fiftieth anniversary of the county fire brigade in Hampshire, held at the Service Headquarters in Eastleigh on 16 August 1998. (Roger Hansford)

This BOC tanker is fully equipped to meet UK regulations for carrying hazardous substances by road. The green diamond on the hazchem plate represents compressed gas, and the product number '1977' denotes liquid nitrogen is being carried. The hazchem code '2T' indicates firefighters would need to use breathing apparatus and dilute any spillage using a fine water spray. Some products would be tackled using a coarse spray, foam, or a dry powder. (Roger Hansford)

Incidents on the Waterside in the 1960s and '70s

Hampshire Fire Service attended a number of serious incidents on the Waterside during the 1960s and early 1970s. These included seven major fires at Fawley Refinery between 1962 and 1969, details of which are described in Alan House's *Proud to Serve* (1998). Normally fire appliances from Hythe and Beaulieu were the first in attendance after Esso's own crews. There were two fires in the catalytic cracking plant, and three in the site's power formers. The most extreme of these incidents required up to fifty water jets, with fifty-seven fire appliances attending and up to ninety-two appliances travelling for cover purposes throughout Hampshire and the neighbouring brigades. Just before midnight on 3 January 1967 there was an explosion and fire aboard the tanker *Esso Glasgow* which was loading petrol and diesel fuel at Fawley's marine terminal. Eleven firemen were injured when a second explosion occurred on the vessel ninety minutes later. Another serious incident occurred in October 1970 when the tankers *Pacific Glory* and *Allegro* collided off St Catherine's Point in the Solent. The resulting catastrophic fire, attended by Hythe and Beaulieu firemen, was put out using foam-spraying tugs.

Not all of the incidents were so large in scale but they still required the Fire Service. Hampshire were called to a fire aboard *Esso Mercia* at the marine terminal on

A more unusual use of the 'Prontosaurus' (see Chapter 5) was at a rail tank fire in 1976, supplementing the fixed firefighting systems at Fawley Refinery's rail loading bays to control a fierce LPG fire. (Alan House Collection)

28 November 1972, and to one at the refinery's rail loading bays on 22 December 1976 where two workers had suffered burn injuries. The *Southern Daily Echo* describes fires during this period at Monsanto Chemicals, Fawley Power Station, International Synthetic Rubber Company (ISR), and Husbands Shipyard in Marchwood. Some of these had their own fire teams, as detailed in Chapter 5. The newspaper reports serious heath fires in or near the area in 1963, 1968, and 1974. On 17 March 1962, crews from Hythe and Beaulieu attended a house fire which was the scene of a double murder and suicide at Mopley Farm, Langley. This was the most macabre of several notable domestic incidents, including a garage explosion in Ipers Bridge and fires in the homes of Esso workers in 1963 and 1967. Fire broke out in the roof of the American army base in Hythe and in the fourth-floor art department at Hardley School in the spring of 1977, requiring the attendance of crews from Hythe, Beaulieu, Lyndhurst, and Brockenhurst. Where these emergencies occurred on land formerly belonging to the Cadland Estate, it is clear how much had changed in Fawley parish since the nineteenth-century fire at Cadland Farm.

3

A NEW FIRE STATION AT FAWLEY: THE HAMPSHIRE FIRE BRIGADE YEARS, 1977–92

The amount of change in the area and the number of serious incidents during the 1960s prompted the building of a local authority fire station for Fawley. As well as better addressing the risks on the Waterside this would improve response times and enable local firemen to train more effectively. The new station was sanctioned by Her Majesty's Chief Inspector of Fire Services and had been discussed by the Fire Service Committee, later to become the Public Protection Committee, as early as 1973. Several different sites were considered: one opposite Fawley School, one at Blackfield crossroads opposite the Royal British Legion, and one in Lime Kiln Lane, Holbury. The site chosen was just south of Hardley Roundabout and at that time adjacent to the Pinewood Transport Café The station was named Fawley as the petrochemical complex was the most significant factor behind its creation. It was to serve an area broadly based on Fawley parish, covering about 35 square miles. The border of this territory ran from Hardley to the Heath Roundabout, to Hilltop, in to Rollestone, down to Lepe, and back around the coast to Hardley. Fawley Fire Station would primarily serve the communities of Hardley, Holbury, Butts Ash, Blackfield, Langley, Lepe, Ashlett Creek, Fawley, and Calshot. The fire brigade during this period was called Hampshire Fire Brigade, formed on 1 April 1974 with the amalgamation of Hampshire Fire Service with the Portsmouth and Southampton Fire Brigades. Fawley was allocated station number '58' which was the highest in Hampshire, and allowed for additional new build stations to be numbered. It was within the New Forest and Southampton region, known to the Brigade as 'D' Division.

Local residents remember seeing the land for the new station being cleared. The tender for building works was dated 12 March 1975 for a cost of £359,983 (see shelf-mark H/CL8/525 at Hampshire Record Office). The contractor was Alfred Lansley of Southampton, the architect was C. Stansfield Smith, and the surveyor was K.W. Bailey. The main accommodation downstairs consisted of a three-bay appliance room and gym,

Fawley Fire Station in May 1978, showing the main building and drill tower. The two appliances, both built by Bedford, are the 1977 water tender ladder OOW 53S and the 1967 foam tanker HOR 853E. (Alan House Collection)

kit room, offices, and a lecture room; there was a bar, kitchen and leisure area upstairs. At the back was a sizeable yard with further garaging and storage facilities on one side and the brick-built five-storey drill tower and smoke chamber at the rear. The initial plans from January 1975 (H/CA2/3/4) showed a station officer's house and space left for expansion next to the appliance bays, but these were not constructed and several trees stand in that location. Two terraces of modern-style houses were built around the station for firemen to live in, all twelve owned by Hampshire Fire Brigade. Because of the houses, and the fact that firemen were employed on the day-manning duty system, there was no need for sleeping accommodation inside the Station. The living area and training facilities made Fawley stand out among fire stations in that part of the county; it was also unusual for a station to have its own access road. Falconer Court was named after Colin Falconer who had served in the Esso Fire Brigade and was in charge of Hythe Fire Station from 1962 to 1976. Initially separated from Harrier Way by a grass bank across the road, it leads out onto Long Lane, the main thoroughfare running south from Hythe known as the A326. Later the grass bank was removed and an electronic barrier placed at the entrance, with signs directing non-emergency traffic to the rear access. There have never been traffic lights to allow fire engines easy access onto the main road, although this has been mooted.

The site was first occupied in May 1977 by Station Officer Malcolm Collier and became operational on 22 June that year. Mr Collier reported to Roger Penny, who was Sub-Divisional Commander for the New Forest, this being a wider area in which Fawley's appliances could provide support to other crews. Local reporter Janice Longland wrote for the *Waterside Observer* in October 1978, saying that Malcolm Collier had been

This was originally the only way in and out of the fire station and its houses, but the exit onto the A326 is now reserved for emergency vehicles on call. An exit for non-emergency traffic was opened when the automatic barrier was installed. (Roger Hansford)

born in Dibden Purlieu and rose up the ranks after joining Hampshire Fire Service in 1966. Promotions had taken him all over the county and he had worked hard for the Fire Services National Benevolent Fund. He also had expertise in fire prevention which became part of his responsibility on the Waterside, including at that time the inspection of premises. With his appointment at Fawley he had returned to his roots. He would be succeeded as Station Officer in 1981 by Frank O'Brien, who was born in Southampton and had served in the city fire stations of Redbridge, the Docks and St Mary's.

An important member of Malcolm's crew had also grown up locally. Chas McGill had served as a retained fireman at Hythe from 1966, and his knowledge of the area was invaluable as Fawley became established. For instance, he knew where all the local hydrants were. Chas had attended his first fire at Fawley Power Station and went to the *Pacific Glory* disaster where he had to climb aboard the burning vessel and encountered fatalities for the first time. He remembers working with Deryck Ayres, Bob Streeter and Colin Killacky, who was Sub-Officer. Eventually there were fifteen full-time crew members working at Fawley, with eight or nine covering each shift. They were on the station during weekdays and Saturday mornings, otherwise responding from home. Chas was one of three who lived off the site in their own properties. He says the fire station was a great place to work, and there were some fantastic characters in the social club. The crews would play volleyball each day to maintain fitness, and they spent time training on local sites to increase their readiness to perform firefighting duties.

Industrial action by fire crews delayed the official opening of the fire station, according to notes made by Fawley historians. This was the first ever national firemen's strike.

The *Advertiser & Times* reported on 26 November 1977 that Fawley's new full-time fire-men were picketing the main entrance road to the Esso Refinery. However, a mock emergency exercise took place in the refinery's rail sidings in the same month with the participation of Fawley's crews and equipment. This may have been in response to the real incident at the same location the previous December. The opening ceremony of Fawley Fire Station was held on 14 September 1978, also the twenty-seventh anniversary of the refinery's opening. The fire station put on a sherry reception and gave a firefighting display using its new vehicles. Hampshire's Chief Fire Officer, G. Clarke, welcomed the 250 guests, and there was a speech by Councillor Norman Best of the Public Protection Committee. Divisional Commander H.G. Stinton was also present. From that September evening in 1978, Fawley Fire Station was officially open and ready to serve its community.

Fawley's Retained Firemen

Chas McGill was made a Sub-Officer in 1981, by which time he had overseen the set-ting up of the retained section at Fawley to work alongside the daytime crews. Malcolm Rumsey joined the fire brigade on 19 May 1980 as fireman no. 3496, and stayed for eleven years; his memories give an insight into the life of a retained fireman in the period. He was working full-time in Southampton and therefore provided night-time cover at Fawley. This took a lot of dedication, as he sometimes had to go to work having been

Fawley firemen training in the drill yard in summer 1981, using the Simonitor vehicle (right). From left to right, the buildings in the background are: the Forest Home public house, the Pinewood Transport Café (behind tower), and the original Hardley School building. (Alan House Collection)

called out overnight, and therefore after little sleep. Sometimes he would sit up and drink coffee in the early hours of the morning whilst trying to process more traumatic aspects of the calls he had attended. Malcolm always kept a photograph of his family with him in his wallet and this reminded him to think of his own safety and that of other crew members when he was on a call. Malcolm said the work of a retained firemen was so demanding it put the men's other jobs at risk. At that time firemen were paid £6.71 for each turnout, or £3.00 'attendance fee' if the appliance was fully crewed or had left before they reached the station.

Malcolm has vivid memories of training to be a fireman. He attended his three-day recruits course at Eastleigh in June 1980. By 19 May 1981 he was qualified as a Category C driver, enabling him to drive Land Rovers. He remembers the test did not include off-road driving. Malcolm obtained his Category B licence on 24 September 1981, which entitled him to drive all fire appliances except turntable ladder or hydraulic platform vehicles, but not on blue lights. He had first needed to gain an HGV licence for rigid trucks, and started by driving around all day with four to six other men, each taking an hour at the controls. With the Category B licence he could drive back from fires but not to them until he became a Red Driver in 1982. His first drive on blue lights was to a fire at Applemore School at which four fire appliances were required. It was rush hour and he recalls driving down the centre white line of the A326 Marchwood Bypass and thinking to himself, 'this is great'. On this occasion the felt roof of the school had been set fire to by workmen and there was a considerable amount of smoke.

Malcolm was appointed Leading Fireman, beating competition from ex-soldiers, when ranks were introduced into the retained section. He had completed his six-month

Bearing interesting comparison with the RTC staged at the fire station open day in 2012 (see Chapter 4), this similar drill took place at an open day in the late 1980s. Malcolm Rumsey is the tall firefighter in the foreground. (Malcolm Rumsey)

probationary period by 1 October 1982. By this time he was also qualified to use breathing apparatus (BA), which was indicated by a yellow diamond inside a black circle on the sides of the helmet. The BA assessment was a particularly difficult one to pass. Malcolm remembers Fawley's first open day on 30 July 1983, one of many the station would hold, with much input also from the whole-time crews. Julia Marsden, aged 15, officially opened this event as 'Miss Waterside', and she was pictured in the local press assisting Leading Fireman Jock Stewart with a hose demonstration. The *Echo* report of the open day said the station had fifteen retained firemen at this point.

Malcolm's interest in training fed into the way he managed the retained firemen, developing their skills. The station held a drill night one evening per week, and records show how this tessellated with the quarterly training schedule provided across 'D' Division. For example, in spring 1985, Fawley's crews did drills for situation, rescue, smoke and heat, lockers, ladders, knots and lines, foam firefighting, and radio communications. In the same month, they also did drills at Esso Fawley and with the Esso Fire Department, and Station Officer Frank O'Brien arranged for them to visit a Very Large Crude Carrier (VLCC) at the Esso marine terminal. Sometimes they trained with crews from other stations in 'D' Division, and regular assessments were held. They also trained with *Fireboat 39*, which had been built for Southampton Fire Brigade in 1963 and remained in service until October 1986. Following the closure of Western Docks Fire Station in September 1984 this was crewed from St Mary's Fire Station. Malcolm remembers that drills in Esso were the least fulfilling, as the crews were simply pumping water for a prolonged period! Malcolm also designed his own training materials and his drill plans were very clearly illustrated. It was, and still is, a source

Two retained firemen from Fawley: Leading Fireman Malcolm Rumsey (left) and Sub-Officer Terri Smith (right). Terri later became chief of the Esso Refinery Fire Department. (Malcolm Rumsey)

of pride to him that his crews were always in readiness and trained to the best possible level of experience and knowledge for what might happen. He also tried to use spare vehicles for training so the water tender was free to go to emergencies at short notice.

Four pagers were kept in Malcolm's house, and when a call came in they would all go off. Malcolm's son, now a retained firefighter at the station, remembers hearing them. As the retained section used to crew the water tender and the special vehicles, they would usually be driving into Falconer Court in their cars just as the day-crewed water tender ladder rushed out. On one occasion the roller door rebounded, pulling the ladder off as the appliance emerged! While a 'trigger 2' was for both fire engines, a 'trigger 3' was just for the water tender, often to attend a domestic incident. Alan House was the Station Commander between Malcolm Collier and Frank O'Brien, and he put the retained crew on call to attend incidents on their own, as they had built up enough experience by that point. The teleprinter gave details such as 'fire – person reported', 'lorry fire', or 'caravan fire', also giving the address and telephone number of the incident and listing the appliances called. Whilst travelling to an incident, Malcolm would use the radio to book 'mobile' and check the address with control, booking the appliance 'in attendance' when they arrived. Sometimes a printout would read 'fire exercise, no blue lights or horns', and these often took place in or near Esso, with the firemen setting out from the station at a pre-arranged time. Malcolm kept a notebook where he recorded the details of incidents to help his own decision-making process, and his notes would be used as a basis for radio messages sent to control.

Malcolm Rumsey wearing the fire kit of the 1980s. Note the tunic with silver buttons. On his helmet, the single black line indicates the rank of Leading Fireman, and the black circle with diamond shows the wearer is BA trained. (Malcolm Rumsey)

When the retained firemen attended calls, these could be very demanding. The uniforms at that time consisted of yellow trousers, black lancer tunics with silver buttons, and cork helmets with no visors. Some firemen found them cold, with insufficient waterproofing, and they were not high-visibility like today's uniforms. Firemen were forbidden from wearing watches or having beards or long hair. They usually wore tracksuits underneath their fire kits for warmth and comfort. This was necessary at one incident Malcolm attended in Cadnam where he needed to climb a ladder and create a fire break in the thatched roof of a burning cottage. When he came back down there were icicles of 2 or 3in in length hanging from his helmet, and the crews' hoses had frozen solid! Another amusing incident was a fire at an old house on Forest Front where potatoes were kept in sacks in the bedrooms; afterwards there were roast potatoes everywhere! On the more serious side, Malcolm had to get his crew out of a fire at Langley Lodge as the building was about to collapse, and he recalls fatalities one night at a road traffic accident near Blackfield involving a travelling rock band. He attended many rural fires in the New Forest and also in local barns, where they used to contain the fire but let it burn as a smoke-damaged crop was of no use to the farmer. Once, in the refinery, he helped two other men carry portable monitors high up a tower to spray cooling water onto a power former fire.

Fawley's Fire Engines

Fawley always had an interesting selection of appliances to serve an area where domestic, rural and petrochemical firefighting were all important; its vehicle allocations reflect its difference in risk profile from the average urban fire station. Fawley was a testing ground for a number of prototype appliances which stand out from the brigade's fleet list. This was partly because of the amount of garaging space there compared to other stations and also because it was crewed by full-time firemen who could trial the vehicles during daytimes. For each vehicle I describe its details and provenance before and after being located at Fawley. New appliances tend to be issued to areas with higher risks or full-time crews, moving to stations with lower risks or retained crews as they get older. Eventually they act in support roles in the county before being sold out of the brigade for scrap, use abroad, or preservation by enthusiasts.

From the 1970s the basic firefighting vehicle in Hampshire was the water tender, which carried a 10.5m ladder, water tank, and hoses. In contrast, the water tender ladder was fitted with a 13.5m ladder and equipment for traffic accident rescue as well as firefighting. Both types are referred to in fire brigade parlance as 'pumps' or pumping appliances, distinguishing them from the 'special' appliances which have particular roles to play at incidents. Most appliances at that time comprised a commercially available truck cab and chassis, with the bodywork and locker construction completed in the brigade's own workshops at Winchester or by a specialist fire appliance builder. In terms of emergency warning systems on fire appliances, the brigade had been fitting blue lights since 1962 and two-tone sirens from 1967. For the author these are quintessential hallmarks of a fire appliance!

At first a number of vehicles were temporarily issued to Fawley, which is not common practice for a new fire station. The water tender ladder LPO 891R was a Dennis DJ which came when the station was first occupied in May 1977. Built in the brigade's own workshops, this was the last example from Hampshire's series of Dennis appliances introduced during the 1970s. This one was moved to Alton in September 1977 and entered the county reserve fleet as a water tender before being sold for preservation in November 1996. Built with a Jaguar petrol engine, the vehicle was converted to a Perkins diesel in July 1982 to increase its durability.

Another appliance at the outset was NYR 24, a Bedford SHZ built for the Home Office. This had been first issued to the Auxiliary Fire Service (AFS) in 1954 as an emergency pump (nicknamed 'Green Goddess') based at Bordon Chemicals, and was purchased by Hampshire Fire Service on the demise of the AFS in 1968 for conversion to a pumping unit. It served mainly at Eastleigh but was transferred to Fawley before its sale in October 1978. The station has always needed a four-wheel drive vehicle for heathland firefighting, and the first two Land Rovers were short-term issues. These were YOR 206 and YOR 207, both built in 1961 on a Land Rover 109 chassis and converted for firefighting in the brigade's workshops. Fawley's temporary issues give a snapshot of the vehicles Hampshire Fire Brigade had built or acquired by the late 1970s, the full history of which is documented by Alan House (1998 & 2000).

The first water tender ladder issued to Fawley long-term was the Bedford TKGS registered OOW 53S. This was delivered new on 22 August 1977 and was to serve at the station for twelve years. Unlike the Dennis DJ, which had a crew cab painted in red, this appliance featured silver metalwork along both sides, as was common in fire engine design at the time. The appliance had a red-painted front and roof, with the word 'FIRE' written in reverse to show in car drivers' rear-view mirrors. The body construction was by HCB-Angus, a fire appliance builder based nearby in Totton. Chas McGill says that, despite its plain appearance, this was the best appliance the station ever had. Its capacity to speed to an emergency was a real boon for the whole-time crew. This Bedford was later converted to a water tender and was issued to Overton fire station on 18 April 1991. It became a spare appliance for Hampshire's 'C' Division on 27 September 1995, and was used for driver training before its sale in August 1997.

Fawley usually had a water tender on station alongside the water tender ladder and the Bedford TKEL registration POU 882M moved there in 1979. This vehicle had been built in the brigade's workshops and was issued new to Bishop's Waltham on 18 December 1973. It is the appliance Malcolm Rumsey remembers driving at Fawley. Its Jaguar petrol engine was very powerful, while the brakes were sluggish, meaning it could be difficult to stop once it was going! In August 1985 the vehicle became a 'C' Division reserve appliance and it was moved to West End Fire Station in February 1987, before its sale in August 1991. In appearance the vehicle was very similar to Fawley's water tender ladder ,and the pair must have made a fine sight going to incidents when a two-pump attendance was required. The water tender ladder would go to road accidents and more serious fires, the water tender to smaller fires, and both would go to property fires when people were reported trapped.

This Bedford TKEL water tender, with Jaguar petrol engine, was originally issued to Bishop's Waltham but was in service at Fawley 1979–85. It was frequently used by the retained section. (Alan House Collection)

Fawley opened with two large special appliances, both suitable for petrochemical firefighting, and kept these into the 1980s. A unique vehicle was the Simonitor built by HCB-Angus on a Dodge K1613 chassis, with booms by Simon Snorkel. Registered LOW 465R, it was new to Fawley in May 1977. This type of vehicle was in favour at the time for high-level firefighting, and was thought to be appropriate for the risks found on the Waterside. It could spray foam onto a petrochemical fire from a height whilst the firemen remained inside, protected from the heat and danger. It could accommodate a full crew and initially was intended to run as the second pump on the station. Chas McGill says it cost nearly £50,000. Although the Simonitor did take part in exercises such as at the Fawley rail sidings in November 1977, it was never actually used for its intended purpose and was taken off the run in May 1983. The unusual story of this fire engine continued when it was converted to a recovery vehicle and placed at the brigade workshops in January 1986, where it remained until being sold in October 1997. The workshop vehicle had the same frontage and registration number but otherwise bore little visual resemblance to the emergency vehicle placed at Fawley when the new station opened.

Fawley's other special appliance was a purpose-built foam tender, HOR 853E. This was a classic-shape Bedford TKEL with body conversion by brigade workshops. It could carry foam in large quantities to the scene of a petroleum fire or spillage, the foam being used to remove oxygen and therefore prevent or extinguish flames. The vehicle had its first run at Basingstoke on 8 September 1967 but was based at Fawley from the station's opening until being sold in July 1988. It has since been in preservation. The wheel hubs, roof, and entire crew cab were painted in red, with the badge on the rear doors. Above the lockers on each side it carried signs saying 'Hampshire Fire Brigade Foam Tanker'. Apart from having no ladders this looked like an ordinary pumping appliance and it was one of two foam tankers built in the same style. Hampshire built around forty vehicles, mostly pumping appliances, on this cab design during the 1960s and '70s, and they did not change the TK cab shape as extensively as commercial fire appliance builders. The series included a canteen van that served in Hampshire until 2000.

Front-line pumping appliances tend to be replaced more frequently than specials, as they are more heavily used and less costly to design. The Bedford TKEL water tender was replaced by a Bedford TKGS, which moved to Fawley on 1 August 1985. This carried the registration number B167 TPX and had a more modern appearance than the earlier Bedfords, particularly with its solid front grill and windscreen. The front displayed the word 'FIRE' in reverse with the Bedford marque given between characteristic double headlamps. This water tender was used particularly by the retained section, and Malcolm Rumsey remembers that its pump was difficult to prime. To operate the power take-off you had to put the vehicle in first gear and then pull a lever under the driver's seat. For firemen travelling to fires it was much easier to put on BA sets in the back of this vehicle than in older vehicles. The Bedford was moved to Hamble Fire Station in Southampton in December 1991, becoming a reserve appliance for 'A' Division in February 1997 before its

Bedford water tender B167 TPX in action at a fire in a derelict house in Heather Road, Blackfield, in 1989. (Alan House Collection)

sale in March 2000. The author remembers seeing this appliance at school open days, and could easily distinguish it from the silver-sided OOW 53S it ran alongside.

The vehicles described above facilitated firefighting in Fawley during most of the 1980s, but changes came towards the end of the decade, not least a new livery of double yellow stripes along the sides of vehicles. A water carrier arrived in January 1988, and this was an essential tool for rural operations. Unlike the foam tanker HOR 853E, the water carrier had a cylindrical tanker body so did not look like an ordinary fire engine; it also had a more antique appearance than the foam vehicle. The tanker could feed the pumping appliances by collecting extra water, sometimes pumping out into an inflatable dam and returning to the water source for another load. The chassis and cab were that of the Leyland Mastiff, the registration number was KTR 891P, and the body conversion was carried out by Fergusson. It was first issued to Lyndhurst in August 1976, where Alan House drove it on its first call. It was based at Fawley from 1988 to 1997, but moved to Fordingbridge for nineteen months before being sold in December 1998. A second vehicle of a similar design was built two years after Fawley's, and this served at Andover and Bordon.

A Land Rover appliance often worked alongside the water carrier for heathland firefighting. The Hythe and Fawley Land Rovers were swapped over in May 1988. Both were built by Hampshire Fire Brigade workshops on a Land Rover 109 chassis. Registration number ROT 689S, was issued to Fawley in November 1979 and ROT 688S to Hythe in October 1980, both with a pump mounted on the front bumper. This could be difficult

One of Hardley's later Land Rovers in use as a control point at Exercise Shannon, Fawley Refinery in May 2012. (Roger Hansford)

to prime and the firemen had to take care not to get their fingers trapped. When the swap took place, ROT 689S was converted to a rear-mounted hose reel instead of a front-mounted pump and issued to Hythe. ROT 688S, which moved to Fawley, was closed in at the rear and carried its hoses in a roof-mounted box. It would remain in service until November 1998.

The Land Rover was the first vehicle to despatch to any incident at Fawley Refinery ,since it would be used for command and control purposes. Radio call signs at the time gave the station number together with a code to differentiate between vehicle types. Thus at Fawley, station designation '58', the water tender ladder was 158, the Simonitor 558, the Land Rover 658, and the water carrier 858, and control was 'HX' for Hampshire. A water tender was prefixed by number 2, but the first water tender on a station had no prefix so Fawley's was simply 58. Together, Fawley's vehicles provided an excellent firefighting resource for the community.

Incidents 1977–92

A study of news reports made by the *Southern Daily Echo* offers a survey of calls attended by the station. In October 1977, crews went with Hythe firemen to put out a kitchen fat-pan fire, and also to a blazing caravan in Lime Kiln Lane. The paper reported a fire in the roof of the football pavilion at Blackfield and Langley Social Club in February 1978, and

The line-up at Fawley, *c.* 1985. Left to right: Bedford water tender ladder, Land Rover, Bedford water tender, Bedford foam tanker, Dodge Simonitor. The Land Rover is likely to be ROT 689S as the photograph pre-dates the swap with Hythe. (Alan House Collection)

a fatality at a house fire in Furzey Close, Blackfield, on 30 August 1978. In the same year, on 1 December, a heathland fire 'raged for two hours' at Dibden Purlieu. At Re-Chem International there was a blaze in a railway truck carrying chemical waste in September 1978, with further incidents there in January and June 1979. On the latter occasion a local resident described five explosions with flying debris, although the company said a small leak had been contained and the plant was continuing to operate.

There was a busy year in 1979. A report of an intense house fire in Beaulieu Road, Dibden Purlieu stated, 'Firemen from Hythe and Fawley pulled down the burning ceiling and dragged smouldering furniture outside'. There were two fires on the top floor of the Heath Guest House on 24 March and a caravan fire in Claypits Lane during the early hours of 26 April. Also in April, firemen took two hours using water and hand beaters to fight a fire in 25 acres of fir and pine forest at Dibden Purlieu. On 7 June, lightning struck the outbuildings of a house in King's Copse Road, making electricity meters in the house smoulder and catch alight. Property fires also occurred at the Old Mill in Holbury in a deep-fat fryer, in a bedsit at the Little Haven Old People's Home in Dibden Purlieu, and in the roof of a bungalow in Ashlett Close, Fawley. Fires at Fawley Power Station were reported on 25 April and 4 October; on the second occasion Hampshire firemen worked alongside the plant's own team on a fire caused by a blowtorch in a chlorine cleansing area.

Eight fire engines were called to the Dreamland Blanket Factory in Shore Road, Hythe at 8 a.m. on 27 January 1980 to a serious fire in the testing laboratory. Twelve of the factory staff were present but all were unharmed. On 18 April, crews from Fawley as well as Hythe, Totton and Southampton attended a fire in the boiler room at Husbands Shipyard in Marchwood, which took an hour to suppress. During summer 1980 there were further fires in caravans in Holbury and on heathland at Dibden Inclosure. In November, Fawley and Beaulieu crews tackled a barn fire at Mopley Farm which destroyed 5 tons of hay. On 9 December there were two unconnected fires at Marchwood Power Station in the space of two hours, one in the main roof and one in an engineering lab, and on 6 January 1981 lagging and scaffolding at Fawley Power Station caught fire.

On 6 February 1981 a burglar set fire to a chair at a house in Copthorne Road, Blackfield, whilst the home-owner was at work, and in the following month firemen tackled a fire in the ruin of the old Holbury Manor House. In May 1981, two serious fires made families homeless at the Heathside Caravan Park in Lime Kiln Lane, Holbury. On 26 July a man using a blowtorch to strip paint off a fascia board at his home accidentally set fire to his roof. Although the family fought the fire with hoses, both engines from Fawley and one from Hythe were required, and the crews took fifteen minutes to control the blaze. In early August a stolen Ford Corsair was set alight in Whitefield Road, Blackfield, and the tug *Accomplice* caught fire during welding work at Husbands Shipyard. A floating dock being used to overhaul the tug *Victoria* was set alight during welding work there on 1 October. The *Echo* reported, 'Fire-fighting teams from all over the Waterside raced to the yard', and the newspaper said firemen wore breathing apparatus and were damping down for three hours. Sadly on 6 October a Hythe man, aged 27, died in a fire at Fawley Refinery when scaffolding collapsed; his friend and co-worker were burnt trying to save him.

As vessels in Southampton Water were loaded for war in the Falklands, another battle was being fought against severe heath fires in the dry weather of spring and summer 1982. On 22 April four crews, including Fawley's, attended a heath fire at Dibden Purlieu, watched by large numbers of schoolchildren. In May, the *Echo* reported the worst forest fires in the area since those of summer 1976. On one occasion firemen from twenty-one stations were working for seven hours to save Burley village from destruction. On 19 August, Hampshire Fire Brigade reported their busiest day of the year; they had attended over 100 calls, including fires in the countryside. Fawley attended domestic fires throughout the year.

Fawley and Hythe firemen were called to Applemore Recreation Centre on 4 April 1983 when an electrical fire burned through a junction box in the switchroom; 200 people were evacuated. On 20 June there was a fire in the gymnasium at Applemore Comprehensive School. Workmen who had been repairing the roof threw compressed gas cylinders clear and tried to fight the fire using hoses and fire extinguishers. Six fire engines were called to the school at 6 p.m., and firemen were battling the flames for an hour. Station Commander Frank O'Brien told the *Echo* there had been a 'severe' fire in the gym and the roof space above. On 19 July there was a fire in the cargo hold of *Cape Island* at no. 3 berth on the Esso marine terminal. This was a 3,900-ton tanker carrying lubricating oil from Fawley

Fawley's whole-time firemen in the mid-1980s. From left to right: back row – Fireman Peter Walley, Sub-Officer Chas McGill, Fireman Eddie Holtham, Fireman John Davenport; front row – Leading Fireman Derek Ayres, Fireman Chris Peckem, Station Officer Frank O'Brien, Leading Fireman Denis Stewart, Fireman Tony Campbell, Fireman Les Kellett. (Hardley Fire Station Collection)

to the Faroe Isles. Hampshire Fire Brigade sent fourteen fire appliances to the scene, and the ship's on-board carbon-dioxide firefighting systems were used. One seaman suffered minor injuries.

The year 1983 was bad for heathland fires. On 14 August eight appliances were called to Fawley Inclosure, and the teams spent four hours dousing and beating a fire covering 7½ acres of Forestry Commission land. The smoke drifted over Southampton. The following day, a fire at Spratts Down on Maldwin Drummond's estate took two hours to control. Arson was suspected, particularly as there had been twelve fires in the area since February. On this occasion the wood and a nature reserve were destroyed. At 11.20 p.m. on 1 September, 4 acres of land in Fawley Plantation were alight for about an hour. Again it was thought to be arson, and school education was considered as a solution to this recurrent problem.

Two properties were consumed by fire towards the end of the year. Fire engines from Fawley, Hythe, and Totton rushed to Tavells Lane, Marchwood, when a gypsy caravan caught fire in November 1983. They had to use cutting equipment to get through the locked entrance gates, and managed to control the fire after forty minutes. Residents had trouble raising the alarm due to the on-site phone being out of action. On 18 December a fire at Gatewood Hill House, Exbury Road, Blackfield required the attendance of crews from Fawley, Hythe, and Beaulieu. They worked for over an hour using jets and breathing apparatus, but were unable to save the house which was partly built from asbestos. The property was empty, but it was thought a cigarette end may have been left there during a teenagers' party that evening.

The potential for serious industrial fires in the area became a reality in 1985. Redundant oil storage tanks at Marchwood Power Station were being demolished using oxyacetylene torches and residual oil inside them caught fire. The situation was further complicated as other workers were removing asbestos cladding from the main power station. At least four appliances were called to the site in February, May and November, and foam was used to extinguish the flames. On 17 June Hampshire Fire Brigade was called fifteen minutes after fire broke out in the engine room of the 3,000-ton Finnish ship *Nestegas*, berthed at Esso. The ship's own firefighting team used four fire extinguishers plus cooling jets to control the blaze, and Hampshire firemen wearing breathing apparatus were required to vent the area. Six fire engines attended the incident.

Just a month later, on 16 July, some fifteen Hampshire appliances rushed to Fawley Refinery after an explosion and fire in the no. 2 power former. Naphtha gas escaped and sent a column of flame 200ft up into the air. A former Esso worker who lived in Holbury raised the alarm immediately. Local residents and the pub owners decided to vacate the Holbury Inn on Long Lane. Local councillor Jack Maynard questioned safety at the plant, particularly as New Forest District Council had that day approved Esso's plans to store LPG on their site. Two days later an *Echo* report announced the refinery would provide an emergency evacuation alarm for local residents. This report also erroneously criticised the age of Hampshire's foam tankers and suggested a fireboat was needed. The new alarm was not sounded on 9 August when five Hampshire appliances attended the refinery after lightning struck a storage tank; three stood by and two provided foam to support Esso's own fire team.

Fawley crews went to serious fires in council houses in Netley View and Calshot during October 1985, and continued to respond to minor domestic incidents on the same estates throughout the following year. There were other, more serious calls. On 7 June 1986, the *Echo* reported: 'Waste chemicals caught fire at Re-Chem factory in Charleston Road, Fawley. Three special foam-filled fire engines – from Fawley, Hythe and Beaulieu – were sent to deal with the fire.' Whilst all three stations were in the first line of response, only Fawley had a foam tanker. On 28 July at, 11.30 a.m., Blackfield First School caught alight. Twelve builders were working there to refurbish the offices and library, and fifty-two children aged 7 to 13 were on a playscheme at the neighbouring Middle School. The *Echo* wrote, 'Children and workmen ran for cover'. Firemen wore breathing apparatus and Hampton Lane was closed due to debris falling from the fire. Another report three days later said it had taken twenty firemen forty-five minutes to control the blaze, and that Station Commander O'Brien was 'suspicious' about the cause of the fire.

Some unusual incidents occurred over the next two years. In the early hours of 21 August 1986, a woman's car hit a kerb on the A326 and crashed through fencing into the compound of gas valve equipment near Hardley Roundabout. This was part of a gas pipeline to the Isle of Wight, but although the car caught alight there was no gas leak or explosion. The road was closed and the woman was admitted to hospital. On 28 July 1987, an *Echo* report said, 'Fawley firemen used a bucket of sea water to dampen planks on Hythe Pier smouldering because of a dropped cigarette butt'. On 11 August, 'Firemen were called to Fawley Power Station today after a propane cylinder caught light in a duct below ground level'. Just after midnight on 5 September, appliances from Fawley and Hythe were called to the Old Pig Farm in Park Lane, Holbury, to a suspected arson attack on a barn. The water carrier from Lyndhurst also attended. The farm's owners had been subject to a series of criminal incidents on their property, and their barn was destroyed along with straw, hay, and farm machinery.

Newspaper reports of incidents during 1988 testified to the successful work of Fawley's crews. On 7 January a fire at an office in the Old Mill Pub in Lime Kiln Lane was put out before the pub became affected. On 15 February a family with their dog and goldfish were saved from a house fire at Holly Close in Netley View, although their possessions were lost. On 5 April a 7-year-old boy visiting Calshot Castle reported a roof fire but was told he had seen the flares at Fawley Refinery. However, a fire was discovered in the roof of a canoe store at the Activities Centre, and staff formed a human chain for passing buckets of water until Fawley firemen arrived and extinguished the flames. A news report from 11 May read, 'Firemen from Hythe and Fawley dashed to Hythe Marina this morning when an electrical fault caused overheating on a luxury yacht, but the problem was quickly dealt with'.

Later in the year, the crews tackled bedroom fires in Thornbury Avenue in Holbury and Newlands Close in Blackfield, and a burning stolen car on Lepe Beach. On 24 October, Fawley and Hythe crews used breathing apparatus whilst fighting a roof fire at a Hythe bungalow. An *Echo* report of 18 November read: 'Fire Limited to Garage. Two fire engines from Fawley prevented blaze spreading to adjoining garages after motorcycle caught fire near Faircross Close at Holbury. One motorcycle from garage was saved, but others were damaged.'

Hampshire firefighters stand-by at an intensive barn fire. They are protecting the surrounding area during a controlled burn of the damaged stock. (Hardley Fire Station Collection)

On 15 December, Fawley accompanied crews from Hythe and Beaulieu to the Dreamland Blanket Factory where fire had started in a fabric machine. The factory's automatic sprinkler system had been triggered. The next evening three appliances were sent to Fawley Power Station but the fire associated with a boiler was already out.

Fawley crews assisted in fighting extensive gorse fires at Holmsley Ridge near Burley in February 1988, but the problem came closer to home on 4 April 1989 when 22 acres of Beaulieu Heath were destroyed. Crews from New Forest stations were supported by the Eastleigh water carrier, and they were working for many hours following the initial report of a fire at 12.39 a.m. Countryside fires were prevalent in 1989, and the concern to prevent arson was publicised by the *Echo* in May. On 1 August, 20 acres at Green Lane in Blackfield caught alight and threatened Mopley Farm buildings. Firemen battled for nearly five hours and a group of up to thirty onlookers blocked the arrival of reinforcement appliances. On 5 August 1989 Hampshire Fire Chiefs called for care to be taken after 133 heathland fires in the area the previous day, including one along the verges of the A326 which had taken an hour to extinguish. Two days later it was reported that 142 fires had been tackled by Hampshire Fire Brigade, including 6 acres of plantation at Denny Wood near Beaulieu Road Station. On 8 August the brigade was still being stretched and arsonists, including in Hythe, were compounding the problem. Arson may have been the cause of fires at Borne Hill Lodge on the Cadland Estate in May and in an

Fawley's Bedford water tender ladder in action at Stanswood Farm, Calshot on 17 August 1979. The crew is using an inflatable dam to attack the fire. (Alan House Collection)

eighteenth-century barn at Little Stanswood Farm near Calshot in September, causing £18,000 worth of damage. Both incidents required the attendance of fire crews for up to three hours. These were minor compared with September's fire at Bratley Arch in the New Forest, which extended over 200 acres, and as a result of which the A31 between Ringwood and Cadnam had to be closed to act as a fire break.

Fawley's firemen continued to attend house fires throughout the year, although sometimes their access to properties was hindered by parked cars. They also extinguished a car fire at Hilltop, where a Ford Orion stolen from Hardley Industrial Estate had been driven and set alight. Two major incidents occurred in the petrochemical complex during 1989, the first at Exxon Chemical on 3 July when fire broke out at 8.30 p.m. in the Butyl Rubber finishing shed, doing up to £50,000 worth of damage. Some fifty firemen were involved, including Exxon's own teams, and thirteen Hampshire Fire Brigade appliances were sent. The *Echo* said that a fire had occurred in the same location fifteen years before, although at that time water sprinklers were not fitted. In the second incident, fifteen fire engines were called to Esso Refinery on 7 November when light crude oil spilled into the engine room of the VLCC *Mobil Petrel* at the marine terminal. Alan House (1998) writes that the effort to avoid an explosion and fire aboard the vessel 'ultimately became the largest use of hi-ex [high-expansion] foam ever in the UK'. He says that weather conditions hindered attempts to remove the vessel to a safer, offshore location, and that, of 127,500 litres of high-expansion foam sourced from across Europe, 86,000 litres were used on the vessel, involving 1,236 staff over fourteen days. Esso Fire Chief Barry Browning commanded the incident and he requested initiation of the Solfire Plan to close the port of Southampton. He recalls that Mobil, Angus, the Fire Service College and Marchwood Army Base all provided assistance with the emergency.

A teenage boy suffered burns to his legs and feet in a bedroom fire at his Furzedale Park home in Hythe on 28 January 1990. Crews from Fawley and Hythe were on the scene within three minutes and, using four sets of breathing apparatus, had the fire under control in eight minutes. On 23 March, controlled burning at Dibden Inclosure spread too far just after 1 p.m. Fawley's crews were joined by others from Hythe, Beaulieu, and Totton, and a Land Rover and a water carrier were used. The fire was quelled before it reached the forest. A Fawley fireman was taken to hospital for smoke inhalation after a fire at Hatchet Pond on 6 April 1990, which may have been started by a discarded cigarette end. It was spreading towards Pilley and Boldre and the fireman became affected at the 'turning point' when the fire was stopped from crossing a Second World War airfield road. Another fire also resulting from controlled burning in that area was reported in October but it was less severe. A pan fire on 8 June caused extensive damage to the upstairs and roof of a preserved 300-year-old cottage in Marchwood. Engines from Fawley, Hythe and Totton attended and crews wore breathing apparatus due to burning asbestos. At midday on 15 June, Fawley and Hythe firemen were called to Malwood Road West in Hythe, where they contained a fire in a flat owned by a couple just celebrating the first anniversary of their marriage. An electrical fire occurred on 9 July at a house in Nash Road in Dibden Purlieu whilst the owners were away in Norwich.

The *Echo* reported a 'Factory Blaze' on 19 March 1991, saying: 'Firemen from Fawley used breathing apparatus to overcome dense smoke when compressor overheated at Crane Engineering Works on Hardley Industrial Estate. Fire was extinguished in 10 minutes.' On 9 May a report detailed recent heathland fires at Badminston Lane near Fawley, Kings Ride in Blackfield, and Lepe Country Park. Throughout 1991 there were also serious property fires. Electricity cables started a roof fire in a bungalow at Rollestone Road in Holbury, reported on 11 February. Three fire engines from Fawley and Hythe attended and the crews had to lift tiles from the roof. The fire was extinguished in ten minutes but salvage operations took an hour. Fawley crews worked alongside those from Hythe and Totton, and the Defence Fire Station at Marchwood Military Port, on 25 March at Cue T's Snooker Hall in Marchwood village. This was an interesting example of collaboration between the Waterside fire brigades. The fire had started in a boiler room and the military police saw the flames at about 5 a.m.. The roof space above was damaged, but fire-resistant materials helped to save the snooker room. Hampshire crews assisted the Esso Refinery Fire Brigade on 4 September after an explosion aboard *Esso Mersey* which was discharging cargo at Fawley. Tragically, two seamen lost their lives in the incident. On 5 September an *Echo* report said an arson attack on Stanswood Mill farmhouse was the second there in three years, and that the attic and roof were damaged. The family was away but would probably have been killed had they been at home. Sadly, a mobile home in Draper's Copse was gutted on 16 December. The couple had just celebrated their golden wedding and returned home from a short visit to Dibden Post Office to find that their home and belongings, including their anniversary presents, had been destroyed in a fierce fire.

The last year of the Hampshire Fire Brigade, 1992, was also the first in which the *Echo* referred to Fawley's teams as 'firefighters' not 'firemen', reflecting the official change in terminology. This is evident in a report of a heath fire on Exbury Road at Blackfield, dated 20 January. Fawley's engines went to a range of other incidents. Both a barn fire at Rollestone Copse on 12 April and a mobile home fire in Church Farm Close at Dibden on 30 April were thought to have been caused by arson. House fires in the area included derelict and occupied properties, and a car fire in Fawley Inclosure caused an access bridge to collapse. On 18 May the second fire in two days at the derelict Dreamland Blanket Factory was reported. The site was awaiting residential development, and juvenile arson was blamed. On 26 May a fire near Dibden destroyed 2 acres of trees and undergrowth. It took an hour to bring under control, and required the attendance of one pump, two Land Rovers, and a water carrier. The Fawley station, along with crews from other New Forest stations, had attended serious fires near Burley earlier in the month. A report dated 12 August 1992 records one of the last calls under the old 'Fire Brigade' title. It said a display caravan parked at Gang Warily Sports Centre near Fawley was destroyed by fire and, 'two crews from the village put out the blaze at 5pm'. It was both the end of the title and the end of an era.

4

NEW NAMES, NEW CHALLENGES: DEVELOPING A MODERN FIRE & RESCUE SERVICE AT HARDLEY, 1992–2014

The 1990s was a period of significant change for the station. Hampshire Fire Brigade was renamed Hampshire Fire & Rescue Service (HF&RS) on 1 September 1992, and Fawley was renamed Hardley in October 1995 following a visit from then Chief Fire Officer Malcolm Eastwood. Deputy Chief Fire Officer Alan House had made the case for the station name to change for geographical reasons, and he implemented the change. It was thought that both new titles better described the community and the way it was served. The station's designation changed from D58 to C58 when the county's four divisions were reorganised to three, and it became part of the New Forest Group once divisional organisation had been discontinued. The comfort and safety of the crews' uniforms had improved from the Nomex fire-resistant clothing of the 1970s to the 'bunker' style tunics of the late 1980s, and Gallet helmets were introduced in 1996.

The years since 2000 have brought increased emphasis on costings, response times and community fire-safety education, including at local schools. More partnership working takes place with personnel from outside agencies, such as Police Community Support Officers, ambulance staff and members of the Maritime & Coastguard Agency. Hardley is a stand-by location for South Central Ambulance Service vehicles, and some firefighters have been trained as community first responders able to deal with medical incidents. Exxon-Mobil Corporation funded a £15,000 Co-Responder Emergency Response Vehicle for the station in 2012, and one of its drivers is an employee of the company. Today there are fourteen members of the retained crew, with three crew managers, working under Station Manager Adrian Butt. The crew is committed to responding to the fire station within five minutes of a call and to turning out within five minutes. Chas McGill followed Terri Smith as Watch Manager in charge of the station in 2010. Chas combines this role with a full-time job in the HF&RS Communications Department, drawing on his training as an electrical engineer, and

Fire appliances at the Hardley Fire Station open day held in September 2012. These vehicles (left to right) represent crews from Hardley (station number 58), Redbridge in Southampton (53) and Beaulieu (49). The Redbridge vehicle is an Aerial Ladder Platform (ALP) (Roger Hansford)

he was awarded the Meritorious Clasp for forty-five years of service on 15 November 2011 at Winchester Guildhall.

Whilst Hardley's firefighters have maintained their professionalism and readiness to attend incidents of all types, the balance of fire cover in the area has been a contentious issue. The late establishment of the station in areas formerly covered by Hythe and Beaulieu was brought into question during the 1990s by a reduction in call frequency. Evidence from the local media shows the fire station came under attack as early as 1988, the year in which Leading Fireman Malcolm Rumsey wrote to Fawley Parish Council about the retained section. The reliability of Fawley's retained crews was questioned in the local press, but Malcolm had the job of planning the cover rotas and says there were always men available to crew. His letter to the clerk of Fawley Parish Council, dated 9 June 1988, read:

> We would like to draw your attention to the fact that the Retained unit at Fawley has been giving fire cover to the area for a period of 8 years, thus providing all the benefits of a two pump station to the Waterside area. During this time Retained personnel have accumulated some 40 years of experience covering a wide variety of hazardous incidents, and we feel that it would be a tremendous loss to the area should this unit be withdrawn … The unit has 5 fully qualified Brigade drivers/pump operators, and 9 men trained in the use of breathing apparatus, and the loss of this expertise would represent a waste of ratepayer's money.

Malcolm went on to say that the unit represented a valuable asset for the area at a relatively low cost, representing 'a small price to pay in an area which has a higher than average fire risk'. This letter followed a report to the council by Eddie Holtham, who was a full-time firefighter at the station and a local councillor. On 17 June 1988, the *Southern Daily Echo* said that Eddie's own job was not under threat from the cuts, but gave his view that a saving of £15,000 from reduced fire cover could be cancelled by the cost of damage from just one fire in the area.

A survey of the archives on the Waterside's fire cover reveals the wide range of opinions and statistics reported by the press, and my discussion here reflects what was reported. An *Echo* article on 11 November 1988 said one fire engine had to be removed from the Waterside as part of a £3 million review of the service. The location for the cut had not been decided, but Chief Fire Officer John Pearson said the Hardley retained section needed to become more reliable otherwise it would be transferred to Hythe within two years. In earlier reports, Hardley's Frank O'Brien had said the station's part-time crew were reliable but responded to fewer incidents than the full-timers, and a councillor had described 'poor relations' between firefighters at Hythe and Hardley. On 4 October 1988, the *Echo* reported a 'high turnover of retained firemen' at Hardley and said their retained pump was 'not available for a major part of 1986/87 because of undermanning'. By November 1991 it appeared Hardley did not receive enough emergency calls for the whole-time and retained crews to serve together. Hardley's average of 300 calls per year, compared with the minimum of 600 normally dealt with by a whole-time station, contradicted its high risk band. Don Bates, secretary of the Retained Firefighters' Union, said on 15 November that it would save £300,000 a year to move Hardley's whole-time crews to other stations, but Ken Cameron of the Fire Brigades' Union said whole-time firefighters trained more, were always available, and had quicker response times. The idea of having whole-time firefighters on duty on the Waterside was part of the original idea behind building the fire station at Falconer Court.

The following day the *Echo* reported the decision of the Public Protection Committee that either Hythe or Hardley's retained pump should be removed. However, the outcome was far from settled, and arguments continued, complicated by the fact that Hythe Fire Station was then under threat of closure. Hardley had a modern fire station that was better designed and situated than Hythe's, and chemical activity in the nearby industries looked set to increase, meriting a high level of cover. In December 1991 the *Echo* said there had been only two major incidents at Fawley Refinery in the previous two years, but the risks in the area included oil trains and petrol and chemical lorries on the A326. Councillor John Coles noted the flight path from Eastleigh Airport over the Waterside but criticised Hardley's retained crew, saying they had not responded to 242 calls in 1988 due to understaffing and, conversely, that the unit was more popular because of the recession. However, Chief Fire Officer John Pearson said the crewing problems had been overcome. On 20 January 1992, George Dawson, a prospective Liberal Democrat candidate and local council planner, wrote to Chief Fire Officer Pearson saying that Home Office Standards for response times in the area's risk band could not be met if cover was reduced.

Three days later Brian Welch said Hardley's crews could protect Hythe's residents. There were suggestions Hythe and Hardley should merge and form a new station called Waterside.

A report on 11 April 1992 headed 'Hythe Fire Station loses part of area' said Hardley would provide the first fire engine responding into Marchwood because of their whole-time crews, and that the same could happen to parts of Dibden Purlieu. The allocation of calls in Marchwood among the Waterside fire stations had been debated before. The *Echo* said the Marchwood military fire appliance could not cover the civilian areas of Marchwood, and that the port would need county fire brigade cover if the military fire station closed. Finally, in a report from August 1992, HM Inspector Mr Robins said Chief Fire Officer Pearson was right that Hythe should have closed. But 11,000 Waterside residents had signed a petition and the Parish Council in Hythe had agreed to spend £28,000 per year in a commendable and understandable effort to keep the fire station there open.

However, the cuts shifted to Hardley, which would lose one fire engine and its whole-time firefighters later in the decade. A proposal by the Public Protection Committee to take one pump away from the station, and have the remaining vehicle crewed by full-time firefighters during the day and part-timers at night and weekends was the subject of an *Echo* report on 14 September 1994. The next day the newspaper said Hythe Fire Station was saved, as councillors had agreed to remove a fire engine from Hardley by the end of the year and introduce a new mixed-duty system. Chief Fire Officer Pearson said he would rather have closed Hythe as the new move would cause the loss of five posts. However, a minimum number of calls had required two pumps from Hardley.

A newspaper report on 19 November 1997 said Fire Chiefs planned to leave only retained staff at Hardley Fire Station, and to reduce the crew at Lyndhurst by two full-time members. On 29 April 1998, the *Echo* said Hardley was to be downgraded and would lose its whole-time firefighters. Eddie Holtham had told a Fawley Parish Council meeting that Hardley had the highest risk area in the county. The station had opened with fifteen whole-time crew, reduced to thirteen in 1994 and eight by 1995. These eight were now to be redeployed and there was a risk Hardley could close if the retained scheme did not work. In a report of 24 August 1998, headlined 'Stalwarts stoking up service war: ex-firemen ignite row', senior fire officers disputed concerns over reduced fire cover raised by retired firefighters from Hardley and Lyndhurst. The next month, Dr Julian Lewis, Conservative MP for New Forest East, said the cuts should not be made at both stations at the same time. In a 5 October report, Hampshire County Councillor Alexis McEvoy also defended the station, saying a forthcoming Home Office review was the last chance to prevent the New Forest being left without whole-time protection, and that Esso's own fire team could not cover homes inside the refinery's risk zone.

Despite the valiant efforts of many, Hardley's fate was sealed by 11 August 1999 when the *Echo* reported that Hardley and Lyndhurst would be re-graded from day-crewed to retained on 7 January 2000. At the same time it was announced that the stations at Titchfield and Twyford would close, while Eastleigh and Waterlooville would be upgraded. The Fire Service homes at Falconer Court would be sold and the firefighters living there would receive financial support to move from or purchase their previously job-related houses. Peter East's report on 12 January 2000 said the station would be crewed by twenty-four retained firefighters. The first and last full-timers held a 'wake', organised by Eddie Holtham, at the Forest Home pub: the original firefighters who attended were

Deryck Ayres, John Davenport, Les Kellett, Dave Fox, and the station's first Sub-Officer, Colin Killacky.

The opinions from these reports reflect the differing motivations of those involved in Hardley's changing levels of cover, and they should not detract from local people's appreciation of the firefighting this station has performed. The reduced number of calls in the area may reflect its isolated coastline location and improvements in industrial safety and domestic fire prevention over the last twenty years. Hardley Fire Station continues to have an essential role in the community.

An Officer's Perspective

Colin Partridge was the Station Commander in charge of Hardley Fire Station from 20 January 1996 until the station became fully retained in January 2000. His comments provide a good insight into life there and they show an officer's point of view. Colin became a firefighter on 1 March 1974, the last month in which Southampton City Fire Brigade was a separate service from Hampshire. He served at Redbridge Fire Station, and the first large fire he attended was at Oakfield School in Totton. He also served whole-time at Copnor and, then Southsea, as well as working in various roles at Divisional Headquarters and Service Headquarters in Eastleigh, before taking up his appointment at Hardley. Today Colin works at Service Headquarters dealing with contingency planning and business continuity, and he finds the operational aspect of his work interesting and varied.

Colin says the period he spent at Hardley was 'the happiest time ever in the service'. He enjoyed being responsible for running the station. The whole-time and retained crews were mainly older men, who had served for at least twenty years; they were respectful and keen, and they 'got on and did the job'. Colin's background with Southampton Fire Brigade had led him to expect a high standard of professionalism, with good discipline and training, and immaculate fire stations and vehicles. He believes this is one of the reasons he was happy at Hardley, as the station always maintained these high standards. Colin also enjoyed his liaison role with the Waterside industries, including Fawley Refinery and the adjacent chemical plants, and he had a good working relationship with all of them.

Part of Colin's role as an officer was the ability to gauge when to get involved in a call and when to leave things to firefighters of lower ranks. He would normally go to industrial incidents, road traffic accidents, and fires where people were reported trapped or missing, but would leave property fires and other incidents to his whole-time Sub-Officer. He would attend calls being handled by the retained crews, or when a Leading Firefighter was in charge, and also fires when reinforcement vehicles were requested on top of the brigade's initial attendance (PDA). The firefighter in charge of an incident wears the 'Incident Commander' vest and this can be passed up the chain of command at the discretion of senior officers as they arrive on scene. Colin had a blue light and two-tone sirens fitted to his car and could attend any emergency either dealt with by his own crews or elsewhere in the county as directed by Fire Control. When covering at weekends, he was

An inspection during the 1990s, showing Hardley's station officer, two Sub-Officers, and Leading Firefighter, and the rest of the crew. They are observed by (left to right) Assistant Chief Officer Bill Bushby, Her Majesty's Inspector Nigel Musselwhite, and Station Officer Frank O'Brien. (Hardley Fire Station Collection)

often the most senior officer in the New Forest area. Colin has worked in fire prevention, and he is a Companion Fellow (and examiner) of the Institution of Fire Engineers, an Incorporated Engineer and a Member of the Chartered Institute of Educational Assessors.

Colin has always lived on the Waterside, and he knew its particular risks, having worked in the accounts department at Monsanto Chemicals in Cadland Road (see Chapter 5) until his mid-twenties. He says Hardley covers the highest petrochemical risk in the county, and possibly in the whole of southern England. The station has dealt with serious petrochemical incidents, including some involving seagoing tankers. The secret for handling these specialised risks is risk awareness, good liaison with local industries, regular familiarisation visits and exercises, realistic training, and detailed planning and preparation for officers and crews. Colin read much about petrochemical fires in order to know the risks and the methods of tackling them. He ensured his crews were properly trained to deal with different scenarios. He respected the knowledge and professionalism of the Esso Refinery Fire Department and established a good working relationship with them. There was an agreement that Hardley would carry out the BA training of Esso firefighters in exchange for the hot fire training of local authority crews at the Esso fire pad. Hot training was essential to familiarise crews of local stations with the noise and intense heat of petrochemical fires. Good preparation and knowledge of the possible risks would give firefighters confidence in dealing with such incidents. Knowledge of the different chemicals could minimise, although not eliminate, the unpredictable nature of these types of fires. A number of retained firefighters who worked at the refinery and adjacent sites provided good cover for local fire stations because of their shift patterns and specialised knowledge. Colin says that, when incidents did occur, the engines from Hythe and Beaulieu would normally arrive to back-up his crews within ten minutes; those from Totton and Redbridge arrived within twenty minutes.

Hardley's Fire Engines

The start of a new era in vehicles was marked in February 1990 when the Volvo water tender ladder, registration G168 UPO, replaced the Bedford OOW 53S. Hampshire had begun using the Scandinavian FL6.14 model the previous year. Although Hampshire built some fire engines themselves, their main bodybuilder was HCB-Angus of Totton, a firm that switched to Volvo for its standard chassis when the Bedford company went out of business. Newly issued to the station, the engine reflected the various name changes as it entered service with Hampshire Fire Brigade at Fawley but ended this allocation labelled for HF&RS at Hardley. Beneath the station name, a plate reading '5000' denoted the extra-capacity water tank fitted to the vehicle, and it also had telescopic scene lighting. Its design featured red hub caps, double yellow stripes down both sides, diagonal yellow and red stripes at the rear, and the word 'FIRE' in reverse on the front. During the period of the local government review, the vehicle carried transfers for 'Hampshire County Council' on the cab doors; this was part of the rebranding which created an opportunity for the brigade to be retitled. Vehicle G168 UPO moved to Lymington in November 1997 and was later sold to the Esso Refinery Fire Department (see Chapter 5).

Older Bedford vehicles continued to run alongside the Volvos, and the replacement water tender at the station was the Bedford TKGS, registered E755 HRV. Completed at brigade workshops, and based around a high-strength cab by HCB-Angus, this was the last Bedford vehicle ordered by the brigade. It was new to Winchester on 27 May 1988, and transferred to Hardley on 16 October 1991. It moved to Beaulieu on 24 May 1995 when Hardley was reduced to a single pump station. On 31 March 2000 it moved to become the third appliance on station at Eastleigh, before going to the HF&RS driver training school that summer. Ostensibly similar in appearance to the previous water tender, the newer example had a more solid front image as the headlights were placed on the bumper, with blue flashing repeater lights either side of the 'BEDFORD' wording on the radiator. The two-tones were fixed prominently at the front, with additional trumpets below the bumpers, which made it sound as though two emergency vehicles were approaching when the sirens blared out of synch! E755 HRV served alongside the G-registration Volvo, and was easily distinguishable because of its more traditional shape and the fact it carried a wooden Bayley ladder in contrast to the silver '135' or 13.5m ladder on the Volvo. It used to follow along the A326 shortly after the Volvo had passed. Both vehicles shared the same livery and underwent the same name changes.

A new livery for Hampshire's fire vehicles came to Hardley in 1997 with P966 JTR, also a Volvo FL6.14 water tender ladder, but this time the firm Saxon had completed the fire appliance conversion. With the new colour scheme the front bumper was white, and grey hubcaps were topped with mudguards, painted black for the front and red for the rear. The Fire & Rescue Service badge was displayed on the rear doors and the striping was green and black, reflecting white in direct light. This striping continued around the front of the vehicle so that it was behind the lettering 'Volvo' and 'Saxon', and the word 'FIRE' was omitted. The advantage of Saxon's build was a longer chassis, and the locker section had a flat top instead of the more traditional rearward slope. This appliance introduced wail sirens for fire vehicles at Hardley – it could produce a rapid or slow undulation of pitch as well as a synthetic two-tone. An American-style silver bull horn gave additional urgent warning for traffic to clear. P966 JTR was issued new to Hardley in May 1997 but moved to Alresford in December 2000, and then to the reserve fleet. It was one of a series of new Volvo appliances, including some designated as water tender ladder rescue appliances. The new striping it carried dated back to J-registered vehicles in the fleet, and was seen at Hythe with their pump K189 MPO, which came to the area in October 1996.

The author recalls on one occasion observing the Leyland Freighter water carrier registered D624 DTR responding onto the Waterside from Eastleigh Fire Station. It became allocated to Hardley in April 1997 when their Leyland Mastiff water carrier reached the end of its time on front-line service. It was based at Eastleigh from July 1987 and then at Basingstoke from August 1996. Before arriving at Hardley the vehicle was refitted at HF&RS workshops with the new livery and new locker and equipment space. It was on station for eighteen months and then moved to Fordingbridge in October 1998. This well-known vehicle was the last water carrier at Hardley as the role then switched back to Lyndhurst, that station being issued with a new Volvo/Angloco Water Carrier, R374 TRV, in June 1998. Also, in relation to firefighting in the New Forest, Hardley acquired a new Land Rover registered R375 TRV in May 1998. This was a Land Rover Defender 110 adapted by HF&RS workshops as a four-wheel-drive light pump, and it remained at the station until August 2012. R375 TRV wore the new livery lines but, like Hardley's last water carrier, still featured the old two-tone siren.

A new pumping appliance came to Hardley in December 2000 – X142 FOR succeeded the P-registered Volvo. This allocation was for much longer, lasting twelve years, and it was later joined in the area by its sister X141 FOR at Totton. Apart from additional fluorescent striping, there was continuity in terms of the vehicle's livery and its audible warning systems, but by 2000 Volvo had remodelled their FL6.14 so that the grill and front bumper were shaped more dramatically. The appliance bodywork was produced by Emergency One, as Saxon Specialist Vehicles had ceased trading in March 2005. The new vehicle had an extra-capacity water tank, but it was a water tender and not the differently equipped water tender ladder traditionally placed at Hardley. That role went instead to Hythe, the station receiving W886 WRV which, as the 100th Volvo fire engine in Hampshire, was named 'Centurion'. In terms of the aesthetic changes evident on these vehicles, more substantial blue light bars had been common since S-registration in the fleet, and these were adapted to include yellow flashing lights to make them more visible to motorists.

Station numbers were also added to the front of fire appliances for at-scene recognition or convoy purposes, so X142 FOR proudly displayed the number '58' in its later years.

The terrorist attacks by extremists in New York in 2001 and London in 2005 prompted UK government investment in fire brigade equipment known as the New Dimensions Programme. A series of MAN TGA vehicles adapted by Marshall SV/Multilift for handling terrorist attacks or large scale emergencies – including chemical, biological, radiological or nuclear (CBRN) incidents – was issued nationwide. These were the first government-issue fire appliances since the AFS vehicles issued 1949–68 to address the threat of armed conflict in the Cold War (see Alan House's book *They Rode Green Fire Engines*, 2002). The high-volume pumping units were first employed at the Buncefield oil terminal disaster, when they were released prior to allocation to fire brigades and crewed by national instructors. Three such vehicles are stationed at Hardley for the High-Volume Pumping team lead by Chas McGill, which also involves firefighters from the Hythe and Ringwood stations. All three vehicles moved to Hardley in 2011 after the appliance bays were refurbished; Deputy Chief Fire Officer Alan House placed the vehicles there to make good use of the garaging space and the retained firefighters based in the vicinity. While WX54 VLF was allocated to Hardley from the start, WX54 VMO moved from Hythe and WX54 VUO from Ringwood. These vehicles carry the yellow and red checked 'Battenburg' design and the words 'Fire & Rescue' on the front. They are not in keeping with Hampshire's

A HVP Unit unloads a pod in the Tactical Holding Area (THA) during Exercise Shannon in May 2012, demonstrating the versatile nature of the New Dimensions vehicles. (Roger Hansford)

livery but have been fitted with Hampshire signs since adoption into the county fleet. The respective pods by Hytrans are pump, hose and equipment boxes for high-volume pumping, with 8,000 litres per minute (lpm) capacity. The allocation shows how Hardley Fire Station has once again accommodated more unusual vehicles.

The front-line fleet was updated in August 2012 with a brand new fire engine and Land Rover. With a completely new shape, the Volvo FLL-240 with crew cab was fitted out by Emergency One as a rescue pump, registered HX12 KXA. This vehicle has a higher equipment specification than a water tender ladder, including a light portable pump, positive pressure ventilation fan, the 'jaws of life', and a compressed-air foam system (CAFS) alongside the main water tank for increased firefighting capability. There are graphics for 'CAFS' on the vehicle and the word 'RESCUE' above the '58' sign at the front, as well as a full-width light bar on the roof. The suspension on the vehicle can be lowered for the easy removal of ladders. The new Land Rover is HX12 KXF, a light, four-wheel-drive pump adapted by Emergency One. The first of a new generation of Land Rovers in Hampshire, it is fitted with extra locker space – a convenient location for displaying the station number. It also has features to increase its visibility in heavy traffic, including high-intensity flashing red lights towards the rear, a full light bar, and the word 'FIRE' in luminous green letters on the bonnet.

A brand new fleet of vehicles for Hardley Fire Station: HX12 KXA and HX12 KXF photographed soon after their arrival in August 2012. (Roger Hansford)

This Bedford TKGS/HCB-Angus water tender ladder was stationed at Fawley from 1977 to 1990. Still praised by the firefighters who crewed it, it was finally sold by the brigade in 1997, after being used as a spare appliance and driver training vehicle. (Hardley Fire Station Collection)

Fawley's Simonitor could run as a second pump on the station but also offered high-level firefighting capability. Built on a Dodge K1613 SH15 chassis, with bodywork by HCB-Angus and booms by Simon Snorkel, the vehicle was based at Fawley from the opening of the station until 1983. It was later converted to a heavy recovery vehicle for the brigade workshops at Winchester. (Alan House Collection)

Fawley's retained firefighters during a drill display at an open day in the late 1980s. As well as the Bedford TKGS water tender with wooden ladder, the shot shows other emergency vehicles redolent of the period: Dodge Power Wagon Rescue Tender from Lyndhurst, Volvo police car and Mercedes-Benz ambulance. (Malcolm Rumsey)

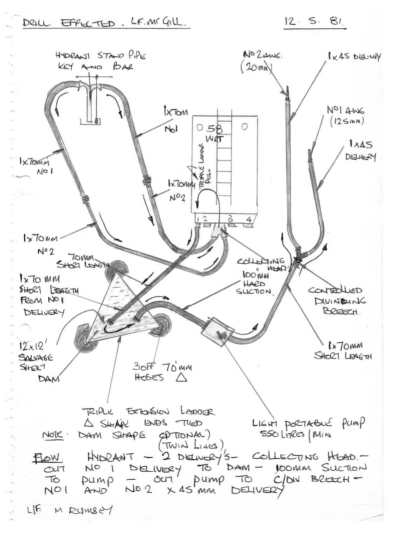

Malcolm Rumsey used to sketch plans for drill nights during the lunch hour at his day job. In this case, the crew must set up two delivery hoses, drawing water from a hydrant, via the fire appliance, and into a dam formed using three ladder extensions with a tarpaulin. Two jets are then run from a light, portable pump. This example shows how the crews were familiarised with equipment and also attests to Malcolm's sketching ability! (Malcolm Rumsey)

An official inspection at Hardley Fire Station during the 1990s. Station Officer Colin Partridge is on the left, with Sub-Officers Terri Smith and Chas McGill on the right. Nigel Musselwhite is the inspecting officer. (Hardley Fire Station Collection)

A line-up of Fawley's appliances (probably before the name change) outside the station in the mid-1990s. Left to right: 1988 Bedford TKGS water tender, 1990 Volvo FL6.14/HCB-Angus water tender ladder, 1976 Leyland Mastiff water carrier, 1978 Land Rover light pump. (Hardley Fire Station Collection)

Hardley has never been without a Land Rover appliance and this photograph captures two generations of the four-wheel drive vehicle. ROT 688S (left) was replaced by R375 TRV (right) in 1998; between them they covered the period 1988–2012. The former was a 1978 Land Rover 109 while the latter was a 1998 Land Rover 110 Defender. (Colin Partridge)

The Volvo FL6.14 water tender X142 FOR, with appliance bodywork by Emergency One, was based at Hardley from 2000 to 2012. The sister appliance X141 FOR was based at the nearby retained station in Totton. (Matthew Leggott)

The High Volume Pumping Unit WX54 VLF, PM 017, has been stationed at Hardley since 2005 as part of the nationwide New Dimensions programme. It can be utilised for industrial firefighting or for flood rescue operations either locally or further afield. (Roger Hansford)

Hardley Fire Station appliance bay, following a refit during 2011 and showing new red doors instead of the original silver ones. Since the refit, Hardley has been the site for all three High Volume Pumping vehicles staffed by crews from Hardley, Hythe, and Ringwood. (Roger Hansford)

Vehicles of Convoy 1, which travelled from Fire Headquarters in Eastleigh to Fawley Refinery for a High Volume Pumping exercise on 5 May 2012, with Dorset Fire & Rescue Service in the lead. (Hampshire Fire & Rescue Service)

KP61 CLU is a Vauxhall Corsa first responder vehicle paid for by the Exxon-Mobil Refinery and crewed by firefighters from Hardley. It is sometimes based inside the petrochemical plant, from where it responds to emergencies in the community. (Roger Hansford)

Hardley's brand new Rescue Pump, HX12 KXA, has CAFS (Compressed Air Foam System) capability as indicated. It has a Volvo FLL-260 crew cab, with bodywork by Emergency One. (Roger Hansford)

E265 JYY began life as a fuel tanker delivering to Esso forecourts from the Avonmouth Terminal. Seen here in the livery of Esso Fire Department, it is carrying fluoroprotein foam and even shows the label 'non-hazardous product' in place of a hazchem plate. The tractor unit could also tow a high-volume diesel trailer pump to where it was needed in the refinery. (Iain Kitchen)

This Leyland DAF ambulance of the refinery's occupational health service carried the personalised registration plate ESO1. It is typical of ambulance design in the 1990s, and displays the logos of Esso Petroleum and Exxon Chemicals. It was sometimes required to attend fire incidents within the petrochemical plant. (Iain Kitchen)

With bodywork by HCB-Angus, the Mercedes-Benz 1936 fire tender G558 DEL was new in 1990. Among the new equipment introduced by Chief Fire Officer Barry Browning, this series of vehicles was based on good practice at other Exxon refineries and it updated provision at Fawley. (Iain Kitchen)

Note the equipment carried by the fire appliance new to Fawley Refinery in 2010. The pumping capacity of the roof-mounted monitors can be increased from 300 to 600gpm using a towable pump, and the monitors can be controlled remotely. (Matthew Leggott)

This Ford A0610 built by Bridge Coachworks was a works fire tender new to Fawley Power Station in 1977, and liveried for the CEGB South Western Region. It was sold for preservation in the mid-1990s. (Roger Mardon)

This Fordson Thames fire tender was featured in a 1962 press article on the ISR Fire Brigade. Noteworthy features are the striped bumper, company logo, equipment strapped to the side, the labelling of lockers, and the industrial setting of the photograph. (Steve Greenaway)

This MAN TGM/ Emergency One water tender ladder was new to Marchwood Sea Mounting Centre in 2011. It is sometimes used as a first response to medical emergencies in the local community. (Roger Hansford)

Three generations of fire appliances at Marchwood Sea Mounting Centre. Left to right: MAN TGM/Emergency One (2011), Volvo FL6.14/ Saxon Volumax (2005), and Volvo FL6.14/ HCB-Angus (1993). The older Volvo was a manual appliance, the newer one automatic, and the latest MAN vehicle is semi-automatic. (Trevor Fenn)

A night-time training drill involving a jet fighter plane, taken at the DF&RS training centre at Manston, Kent. The firefighter to the left is wearing a reflective 'proximity' suit akin to those used in the oil industry to enable close access to fierce fires. For health and safety reasons, the DF&RS is the only brigade still to undertake hot fire training. (Trevor Fenn)

Incidents 1992–2014

An *Echo* report on 1 September 1992, the first day of the HF&RS, said Hardley's water carrier had been sent to help crews from Burley, Brockenhurst and New Milton with a fire in a New Milton bungalow. The vehicle was also called upon the following month, on 9 October, to support two Lymington crews at a hay fire, and on 14 October to a barn fire in Fordingbridge that required six pumps, two Land Rovers, one control unit and one emergency tender. A ship repair workshop near Hythe's Hotspur House was damaged by fire on 16 October, the incident requiring two fire appliances. Hardley's firefighters saved a building at the former Marchwood Power Station on 16 September when a fire broke out in the roof space at about 5.30 p.m. They were called to the site again at 9 a.m. on 20 January 1993 when oil-coated lagging caught fire as workmen were using cutting equipment to demolish the old transformer. Hythe and Hardley crews used foam and high-pressure fog and had the blaze out within twenty-five minutes. The newspaper report of 21 January showed Station Officer Colin Partridge as the senior officer in the photograph, and said that a special foam appliance had been used. Hardley had also been busy on 16 January when a robber crashed a car into a lamp-post near the Circle K store on the A326 and then set fire to it after the police and the car's owners had been to view the scene. On the evening of 16 March, two Hardley crews took ten minutes to extinguish a fire that had spread from a shed into an adjoining house. The owner believed it was arson and he put up a £500 reward for information. After a caravan fire in April on Drapers' Copse estate off Claypits Lane, fire crews from Hardley and Hythe were 'winning praise for their speedy response'. The couple were rescued from their home, and a neighbour saved their Alsatian dog. On 6 June a fire in one of Marchwood's oldest barns required the attendance of fifteen firefighters from Hardley and Hythe. Two BMX bikes were found at the scene and passed to Hythe Police, as they may have belonged to teenage arsonists.

There were some serious fires in 1994. Hardley worked alongside crews from Totton, Redbridge, Lyndhurst and Eastleigh to tackle a fire in a storage tank at the South Western Tar Distillery on 17 February. The site, which had its own fire team and Land Rover, held tarmac and chemicals and this was the second tank explosion there within a few years. Around thirty firefighters were called to a heathland blaze extending over 12 acres at Blackfield in March. The *Echo* report of 14 March read: 'Overhead power cables and poor access made the fire difficult to put out as strong winds carried the flames across the gorse between Badminston Lane and Mopley.' On 15 August, Hardley firefighter Eddie Holtham was walking his dog in Fawley Inclosure and reported a grass fire that was dealt with by his colleagues from the fire station. Another fire was spotted by a teenage passer-by on 5 October who saw flames on the roof of Holbury Junior School at around 8 p.m. and raised the alarm. He was a former pupil of the school who wanted to be a firefighter, but was initially arrested by police on suspicion of arson as he was on the roof trying to tackle the flames when Hardley's crews arrived. The fire was under control within fifteen minutes. Hythe and Hardley were called to an arson attack on a couple's home at Cormorant Drive, Hythe in October. Accelerant had been poured through the letterbox, but a neighbour who had done a fire safety course quelled the flames using a hose.

Also that month was a fire in a small roof space at the US Army base in Shore Road, Hythe. A fire on a landing craft in the River Test near Marchwood Military Port occurred in November on the day when the Princess Royal was visiting. Hardley attended along with crews from Southampton, and a military firefighting vessel was called. However, the royal visit proceeded smoothly and the princess was not made aware of the incident.

During the early part of 1995, firefighters were called to a fire at the former Pentagon Stores in Blackfield Road, and then to an arson attack in Rollestone Road on a converted coach belonging to a well-known rally cross driver. On 15 June, crews tackled a fire at the derelict Westcliffe Hall Hotel which destroyed 25 per cent of the roof. This was the latest in a series of arson attacks at the property over the previous year. About ten days later, 8 acres of farmland were incinerated at Marchwood following a spark from a builder's bonfire at Tavells Farm. A total of thirty-five firefighters worked for several hours to stop the blaze spreading to the Fawley oil tank railway line. During the night of 4 August there were three fires on heathland close to Fawley, one covering up to 10 acres. Residents from Green Lane and Saxon Road were evacuated. HF&RS reported 12 August as their busiest day of the year; crews had attended 177 incidents across the county by 10 p.m., many of which were grass and heath fires. Alan House (1998) notes that twelve emergency Green Goddess pumps were requested from the Home Office to provide back-up for rural locations. On 22 August, water was transported from the Esso Refinery to fight huge fires beside the M27 near Chilworth. Resources were also stretched locally: a report the previous day said Hardley's fire engine had been unable to attend a beach hut fire at Calshot due to being tied up at a grass fire near Dark Lane, Blackfield. This shows the immediate effect of the removal of Hardley's second engine the previous year; the Hythe appliance went to Calshot along with Hardley's Land Rover.

In October 1995, Hardley firefighters attended two incidents involving vessels: an engine-room fire on the 10m tug *Jane Lee* in the Solent and an explosion aboard a £10,000 high-speed fishing trawler moored at Hythe Marina. On the first occasion they were transported to the scene along with Hythe's crews aboard the Calshot lifeboat, and returned after two hours on the fishing boat *Determination*. The firefighting tug *Thrax* from the Fawley marine terminal also responded. Following a barn fire at Heywood Farm, Boldre, on 30 October, the *Echo* reported: 'Crews from Beaulieu, Brockenhurst, and Lymington were joined by colleagues from Hardley Fire Station, which until recently was known as Fawley.' The fire had started just after 11 p.m., but relief crews from Woolston were on the scene until 7 a.m. the next morning. At 9.15 p.m. on 19 November, Hardley firefighters were called to a mobile home explosion on the Lime Kiln Lane Estate whilst many local residents, the author included, were watching *London's Burning*. Another caravan, in Holbury's Southbourne Avenue, was destroyed in an arson attack later in the month. The father of the caravan owner told the *Echo*: 'The fire brigade were here pretty quickly but there was nothing they could do.'

There were further serious property fires at the start of the following year. On 8 January 1996 an arsonist threw a petrol bomb through a window at Applemore College, causing a fire in the creative arts and languages block after which the youngest year groups had to stay at home. Arsonists struck again on 21 February at Priestcroft Pensioner's Home in

Raising the '135' or 13.5m ladder into position to perform a rescue from the drill tower at Hardley Fire Station in September 2012. This drill prepares crews for dealing with fires in property, where people could be trapped on upper floors. (Roger Hansford)

Hampton Lane, Blackfield. They had gained entry to the lounge and started a fire there at 1.20 a.m., just ninety minutes after the end of a concert for residents. A report dated 15 March was headlined 'Crew praised for saving home', and said that Hardley's firefighters had worked alongside those from Beaulieu, Hythe, and Lymington to save a thatched cottage on Main Road in East Boldre which caught fire just after 5 p.m.; damage was confined to the roof. On 14 May the seventeenth-century Staplewood Farmhouse in Marchwood was ablaze, requiring the attendance of fifty firefighters. The homeowners were saved but the water supply was problematic as the nearest hydrant lay a mile away. At 4.30 p.m. on 25 June, twenty firefighters were called to a fire on Beaulieu Heath from the stations at Hardley, Hythe, Beaulieu, Lyndhurst and Romsey. This followed heath fires in the area during April and May. On this occasion the Hardley water carrier brought in 27,000 litres of water, and Lyndhurst's multi-role vehicle (MRV) was also used. Station Officer Colin Partridge said the firefighting operation had contained the fire to 15 acres and saved important wildlife areas. More unusual was a call on 17 September to a peat fire on the banks of the Beaulieu River. The fire was fought from the Harbour Master's barge using a mobile pump and water from the river. In the autumn there were vehicle fires in Blackfield and Butts Ash, and the offices of an advertising agency at Whitefield Farm near Lepe were destroyed with £50,000 worth of damage caused.

In March 1997, arsonists smuggled a smoke bomb into the cereal packet shelves of the Pioneer Store on Long Lane in Holbury. Staff dealt with the resulting very small fire and the shop was not evacuated. On 18 April six fire stations from the Waterside and the New

Forest were called to a fire in the thatched roof of a property on Main Road, East Boldre. Soon after this came a severe forest fire covering 150 acres at Markway Inclosure between Lyndhurst and New Milton, requiring the attendance of 100 firefighters. Two water carriers were sent to back-up the fifteen pumps and eight Land Rovers on the scene, as there were no rivers or fire hydrants nearby. The smoke column could be seen from Hythe and Dibden and it created visibility problems on the A35. The cause was thought to be arson, and a description of the suspect was released on 22 April. Also in April, 33sq. metres of Forest Front Nature Reserve were consumed by fire, and in May a blazing van was discovered on the former airfield at Beaulieu Heath. Firefighters stood by at Fawley Refinery on 24 June while Esso firefighters tackled a blazing oil leak near the site's steam cracking plant. In July firefighters dealt with a car fire in a garage on Cadland Park Estate Road plus a fire at the end of Hythe Pier which damaged the door of a waiting room for the Hythe Ferry. During the early hours of 8 August, twenty firefighters attended a barn fire at Moulands Farm, near Winsor, Totton; they managed to contain it but 500 tons of straw was lost. A second barn fire near Fordingbridge on the same night made both fires appear suspicious. On 15 September, three fire crews, with some members using breathing apparatus, worked for twenty minutes to fight a bedroom fire at Cosworth Drive in Dibden Purlieu. Equipment for Racal Antennas was destroyed by fire on 15 October in a building between Fawley and Beaulieu, probably caused by arson. There were also suspicious circumstances behind a fire at Enichem Sports and Social Club at Rollestone Road in Holbury on 2 November. Hardley, Hythe, Beaulieu, and Brockenhurst firefighters saved the main hall of the building.

Firefighters attended a kitchen fire on the Hollybank Estate in January 1998, but dealt with a much more serious house fire on 14 May that year when a 30-year-old mother of two was killed. Her body had to be recovered by crews wearing breathing apparatus. The fire in the upper part of a house in Ash Close, Netley View, was reported by neighbours at 12.55 a.m. Police began a murder investigation and a 30-year-old man was taken into custody. The burnt-out house was not bulldozed until several years later, which was upsetting for friends and neighbours. In a similar incident on 2 August, PCs Mark Blake and Dave Spelling entered a burning house in Heather Road, Blackfield at around 10 p.m. to save a 36-year-old woman. She was briefly hospitalised, and the police officers were praised by the Fire Service. On 20 July, firefighters from Hardley and Hythe were met by the Special Equipment Unit (SEU) from St Mary's at Hythe Marina due to a report that an incoming boat was on fire. In fact the vessel's engine had overheated and it had given off smoke whilst in the Solent. Another false alarm was reported by the *Echo* on 20 October. Four Waterside fire crews along with the SEU were called to a petrol tanker fire on the A326 near Dibden Purlieu. However, it was simply an overheated wheel bearing. Esso said the vehicle was not connected with their operations. An article dated 24 October read, 'Firefighters were called to a blazing barn at the Applemore Riding stables in Hardley … in the early hours of this morning. Four pumps cleared the area and protected nearby buildings while the flames died down.' Early in December a couple were made homeless at the former Hythe Brewery in Langdown Lawn after suffering two fires in three days. A spark from the first fire on the upper floor must have remained and caused a second one that damaged the roof. The year closed with a blaze in a Dutch barn at Farm Cottage in West Street, Hythe, at which 80 tons of hay was lost. The family moved cows from a nearby field and used their tractor to move burning hay bales out of the barn to stop them causing structural damage. Hythe firefighters moved four oxyacetylene cylinders out to stop them exploding. Hardley and Beaulieu were also at the scene, where muddy conditions made the operation difficult.

The Special Equipment Unit (SEU) from St Mary's frequently attended calls at the Fawley petrochemical complex. The Volvo FL6-H220 appliance was converted by Emergency One in 2006. The allocations for special appliances in Southampton switched shortly before this book went to press, so the SEU moved to Redbridge, and a new ALP was placed at St Mary's. (Roger Hansford)

The final year of the old millennium brought new concerns for the Fire & Rescue Service, including fears of a 'millennium bug' affecting fire alarms and Fire Service computers, and the first threat of strike action by firefighters since 1977. However, the range of incidents attended was as wide as ever. In February, a gentleman awoke in his Stanswood Road home to find his electric blanket on fire. Fire crews from Hythe and Hardley needed two hose reels to tackle the blaze, and Leading Firefighter Deryck Ayres rescued the man's 8-year-old labrador. A Hampshire County Council report from 1998 said 45 per cent of electric blankets were unsafe. A suspicious fire on 28 March 1999 caused £30,000 worth of damage at Ower Farm in Calshot Road, Fawley. Crews from Hardley, Hythe, Totton and St Mary's were called at 4 p.m., and the SEU from St Mary's was requested so access could be gained to the building. The *Echo* reported, 'Flames ten feet high leapt into the air as the intense fire took hold of the barn, packed with hundreds of bales of straw … A huge cloud of black smoke hung over the area and could be seen miles away'. On 22 April, from just before 3 a.m., ten Hampshire fire crews worked for three hours to help MOD firefighters with a store fire at Marchwood Military Port. An MOD forklift was used to gain access as the doors had been buckled by heat. A plume of smoke drifted across Southampton Water but the fire was stopped before it reached the paint store and a gas cylinder was removed from its path.

An *Echo* report on 9 August 1999 entitled 'Firefighters absent from top summer shows' said crews would no longer be standing by at the New Forest Show, Middle Wallop Air Display or Portsmouth Navy Day but would be at Farnborough Air Show and would provide equipment for the Southampton Boat Show. Brigade spokesman David Askew mentioned Hardley and said any of the New Forest stations plus Totton could easily reach the New Forest showground using blue lights. Hardley was also highlighted in relation to a series of arson attacks on their patch in August; the entry read, 'The succession of calls to the Hardley Fire Station follows a Home Office decision to downgrade it by removing its full-time firemen', and Alexis McEvoy said the service was being overstretched. Arson was still a problem in October, causing the station to attend several incidents at which the Hampshire Police spotter plane was used to help locate pockets of heat and to search for the offenders. On 6 November, 75mph 'tornado' winds sent a couple's caravan tumbling 40ft across the beach at Calshot. The man was uninjured but his wife had to be rescued by firefighters and taken to Southampton General Hospital. The woman was later discharged, but quite a bit of damage had been done at Calshot.

On 31 January 2000, Hardley, along with Hythe, attended a fire caused by a gas cooker at a bungalow in Dibden Lodge Close, Hythe. Although the cooker was destroyed, the elderly female occupant was unharmed. A boat caught fire at the Esso Sailing Club at Ashlett Creek in Fawley on 6 April, and the same two stations attended. On 28 April a 3m leylandii hedge was set alight, possibly using an accelerant, in Oakley Close, Holbury. The *Echo* said this was the second arson attack at the property in two months and that, 'the blaze was tackled by a part-time crew from Hardley'. Fire service communications were being discussed in 2000 as on 24 May the paper mentioned that new radios were needed to ensure equipment was intrinsically safe for use on sites like Fawley Power Station. Fire safety came up towards the end of the year as the service had reached the target of issuing all 9,000 Waterside children from Calshot to Calmore with smoke alarms for their own bedrooms. The local schools held non-uniform days to raise money for the alarms, and the Hardley crew raised £3,500 for the project, known as 'Target 2000'.

There was a massive heath fire at Fawley Inclosure on 27 June 2001, in which 6 acres of land were devastated, and thirty children almost became caught in the fire. Some forty firefighters were at the scene, with eleven fire appliances, and the conflagration took eighty minutes to bring under control. Sub-Officer Terri Smith told the *Echo* that the fire was very serious. On 16 July there was a call to a fire at the Flying Boat Inn at Calshot at 9 a.m., attended by Hardley, Hythe, Beaulieu and Totton. This part of the former RAF base had been converted into a pub and restaurant and was unoccupied at the time of the fire. A few days before the incident, an application to turn the area into social housing was rejected by New Forest District Council, as was a larger application in 2005 which would have increased the population needing cover from the fire brigade. Three people escaped alive from an intense blaze at White Cottage, Main Road in Marchwood at around 1 a.m. on 16 August. Sixty firefighters from the Waterside, Southampton, Eastleigh and Winchester were fighting the fire in the thatched property for three hours, using four hose reels and breathing apparatus, and Main Road was shut until 9 a.m. It is thought the residents escaped the fire, which may have been arson, because of their Alsatian dog's barking. A teenage girl survived a fire caused by an unattended chip pan at Tristan Close in Calshot when her parents heard the smoke alarm from the garden, said a *New Forest Post* report on 11 October.

Fires in buildings continued into 2002. On 16 January, Waterside fire crews worked for two hours to deal with a fire in Marchwood's 50-year-old Gospel Church that may have

Hardley's Rescue Pump seen during a fire training exercise. (Roger Hansford)

been started by an electrical fault. A *New Forest Post* report on 24 January described how a 12-year-old boy survived a fire at Eastcot Close in Holbury by jumping from the first floor into the arms of a neighbour. The neighbour also went inside the house to ensure the boy's parents escaped the fire, which may have been started by a cigarette lighter catching on bedding. An *Echo* report of 16 October said seven Calshot beach huts had caught fire the day before, possibly in suspicious circumstances, with the flames being fanned by 55mph winds. A Hampshire County Council highways supervisor had raised the alarm, and a Cadland Estate worker pulled a beach hut out of the way to create a fire break. Crews from Hythe, Hardley, Beaulieu, Redbridge, Totton and St Mary's attended along with Lyndhurst's water carrier.

A varied range of incidents occurred in the area in 2003, some of which were tragic in nature. The A326 was closed between the Heath and Hardley Roundabouts for an hour on Monday 21 April. Twenty-four firefighters from Hardley, Hythe, and Beaulieu reached the scene at about 10 a.m. This was one of a series of fires in the Fawley Inclosure at this time. On 11 June, a couple were taken to hospital with smoke inhalation after a fire in a bungalow at Ashdown Road in Fawley. Their home was severely damaged but the fire was out ninety minutes after the call to HF&RS. On 15 June, at around 5 p.m., a family became stuck in mud close to Hythe Sailing Club after their dinghy ran aground. Hardley and Hythe attended along with the MRV, which by this point had been transferred to Eastleigh. The stranded people were brought back two at a time across special mud mats. An *Echo* report on 15 July said fire crews had given 5-year-old Amie Dixon a Baby Benjamin doll as her family's house had been destroyed and her dog killed at their bungalow in Ashdown Road, Fawley. The gift was organised by Sub-Officers Terri Smith, Phil Gittings, and Charlie Knight. Early in September, there were two fatalities in road accidents close to the fire station, one of which tragically involved a 21-year-old member of Hardley's crew, and this was attended by his colleagues. On 28 September, crews from Hardley, Hythe, and Beaulieu were called to a fire at the Queen Elizabeth II Sports Hall in Blackfield; this caused £5,000 worth of equipment to be lost, and affected several sports groups. Mains gas and power had to be turned off before the fire could be tackled. The *Echo* said, 'the building burned down before the fire could be controlled'. In mid October three quarters of the Forest Front Nature Reserve for disabled people, at Butts Ash, Hythe, was devastated by an arson attack. Work to establish the nature reserve had begun there in 1983. Children aged 8 or 9, thought to have started the fire, were still running around the scene with petrol cans after appliances from Hythe, Hardley, Beaulieu and Lyndhurst had arrived.

A major alarm was raised just after 1 p.m. on 26 August 2004 due to a fire in an oil processing unit at Fawley Refinery. Fire engines from Hardley, Hythe, Beaulieu, Totton, New Milton, Eastleigh and Southampton were mobilised. Vehicles were moved across the New Forest for cover purposes and some stood by at Hardley Fire Station. About fifty Hampshire firefighters supported Esso's own fire team. In total, seven pumping appliances responded to the scene along with an SEU, and the MRV and incident command unit (ICU) from Eastleigh. Police and ambulance staff were also involved. The fire took an hour to deal with but most units were on the scene until about 5 p.m. This was arguably the largest attendance of fire vehicles into the area that year, but there were other serious incidents. On 16 August a wooden building at Blackfield and Langley Social Club 'was burnt to the ground' in a fire, possibly caused by arson, and this was attended by thirty firefighters who worked at the scene for three hours. On 7 September a family escaped a fire in a flat above the Old Mill Inn, Lime Kiln Lane, Holbury at around 3 a.m. when they were woken by their smoke alarm. Redbridge firefighters attended to support crews from the Waterside fire stations. A 49-year-old man was taken to hospital with smoke inhalation after a house fire in Long Lane Close, Holbury, on 19 December. His wife and 16-year-old daughter were not at home; three fire crews attended the incident.

Amid fears of contamination from toxic chemicals or asbestos, seventy firefighters worked for eighteen hours in mid-January 2005 on a fire at Marchwood Waste Transfer Station. The sight of police officers wearing gas masks caused panic in the local community, although it was later found there was no risk to the public. The fire was in 200 tonnes of non-recyclable household waste. An aerial ladder platform (ALP) from Dorset was called in to assist the night-time operations, as the ALP at Redbridge in Southampton was temporarily unavailable. A report on 2 June said a sailor had been killed in an accident aboard the vessel *Murmansk* as it prepared to leave the Fawley marine terminal at 6.30 p.m. the previous Sunday. On 9 June, Chris Yandell wrote in the *Echo* that education programmes had reduced the impact of arson in the area. Firefighters had been called to more than 100 grass and heathland fires during 2003 in Fawley, Forest Front, Netley View, and Dibden, but just twenty-eight in the previous year. Hardley Fire Station held a special programme in March 2005 for selected teenagers in the local community, giving them the chance to experience life as a firefighter. The aim was to teach young people to behave appropriately in society, and to learn about fire safety and the dangers of deliberately starting fires. The programme was known as LIFE – Local Intervention Fire Education – and it concluded with a pass-out parade. The teenagers trained using an appliance dedicated to fire prevention – in fact one of the earliest Volvos that operated in the county, which had been retired from front-line duties.

On 4 February 2006, a 31-year-old horse had to be put to sleep after being caught in an electric fence at Calshot. Fire crews from Hardley and Eastleigh worked with a vet but were unable to save the animal; the MRV from Eastleigh also attended the call. A *New Forest Post* report of 23 March described an arson attack at a £1 million tide mill at Palace Lane in Beaulieu. Unique in the country, the Grade II listed building was 400 years old and was in the process of being restored. Some £60,000 worth of wine and cognac stored on the premises was also destroyed. This followed a spate of arson in the area, and the LIFE course was run again at Hardley Fire Station in April. *Echo* reporter Matt Smith said on 3 April that the aim was to reduce arson by 10 per cent across the county by 2010, and that previous courses had proved to be highly successful in changing young people's attitudes. The report also stated: 'A serial arsonist is thought to have been responsible for at least eight fires in the New Forest since November.' Not only did arson increase Hardley's call rate considerably, but it put lives at risk if their vehicles were tied up at deliberate fires when other calls came in.

On 24 June 2006, Hardley and Hythe went to a fire in a 42ft motor cruiser moored at Hythe Marina village, probably caused by an electrical fault. The owner tried to save the vessel, which was worth £250,000. Unfortunately the focus switched to arson again later in the summer. An *Echo* report by Chris Yandell on 19 August said arsonists had started 100 fires between May and July on Forest Front, and in Fawley and Dibden Inclosures. The danger of fire was increased as a hot summer had followed the driest winter for 100 years, and it was a threat to the habitat of the rare Dartford Warbler and other wildlife within the New Forest Site of Special Scientific Interest. On 25 August, fifty firefighters were called to Badminston Common, near Fawley, soon after 3.30 p.m., where a fire covered 100 sq. metres. The attendance included six pumps, two Land Rovers, two water carriers, a catering vehicle, and the ICU. A report on 30 August by Matt Smith said that 150 arson attacks at Forest Front and Fawley Inclosure since May had cost taxpayers £250,000. Firefighters were visiting homes to try to find out information about the suspects. However, there had been just one forest fire in the danger area by December; by this point CCTV cameras had been fitted to the appliances at Hardley, Hythe, and Beaulieu. Hardley attended various types of incident in the latter part of the year: a house fire in Beech Crescent, Netley View on 9 October, a barn fire in Sowley on 20 November, a fire in an industrial rubber-drying oven on 28 November, and a crashed car on the A326 on 29 December. On each occasion they were working with crews from other nearby stations. The fire at Netley View took one hour to control, but luckily a smoke alarm had alerted the 20-year-old mum that her toddler was trapped by a fire in the lounge. The incident on the A326 happened near Applemore Hill when a car veered off the road soon after 1.30 p.m. and came to rest vertically against a tree.

A dramatic fire broke out in the technology block at Hardley School & Sixth Form at around 8 p.m. on Sunday 11 February 2007. Forty firefighters attended from Hardley,

Hardley firefighters remove the roof from a 'crashed' car in the fire station drill yard so that 'casualties' can be rescued. With the A326 passing close to the station, this training is essential for the crew. (Roger Hansford)

Hythe, Beaulieu, Totton, Redbridge, Eastleigh, and St Mary's. The fire took thirty minutes to bring under control and one hour to damp down. Watch Manager Terri Smith said the crews used breathing apparatus and also prevented the fire from spreading into the main school building, although some school equipment and coursework had been lost. Hampshire Constabulary launched Operation Templar to investigate the attack as the same block had been broken into the night before and arson was suspected.

Serious flooding affected parts of the country in the summer of 2007, and the HVP team from Hardley, Hythe and Ringwood was called into action, firstly to assist with flooding in Doncaster in June, and then to Gloucestershire in July. This second time the task was to prevent flood waters rising at a main electricity sub-station in order to prevent a blackout in the area. An *Echo* report on 24 July said the vehicles could pump three times quicker than conventional fire engines, and that – of forty-six HVPs in the country – the Hampshire vehicle was the only one crewed by retained firefighters.

Fires at industrial premises dominated the work of the fire brigade in the second half of the year. Hampshire crews were not summoned to a fire in a Fawley Refinery steam turbine at 7.30 a.m. on 25 July, but the effects were visible from outside the plant, with a smoke plume covering Southampton Water for many hours. Esso's own fire team dealt with the incident, which shut down part of the refinery. Firefighters from Hardley, along with Hythe, Beaulieu, and Lyndhurst, were called to Fawley Power Station on 31 July, where they set up a 200m exclusion zone because of a fire involving oxyacetylene welding sets. After a gas leak at 10.30 a.m. on 19 October, police had to close the A326 and warn residents and schoolchildren on both sides of Southampton Water to stay inside. Although a nauseating smell was released, the site evacuating alarm at Fawley was not sounded. On 30 October, a fire was discovered at 7.30 a.m. at Hythe Boat Yard. An eyewitness said the flames grew to twice the height of the building. The ex-Naval vessel *Medusa*, one of just fifty-eight vessels in the Core Collection of UK Historic Ships, was being restored, and although the ship was unharmed, its engines were lost in the fire. Incident Commander Dan Tasker said there were seven pumps, an ALP, and an SEU on the scene, and that the building was almost totally destroyed. One hundred firefighters were involved in the job, which was complicated by there being asbestos in the cladding of the building. Once the flames had been put out and the area damped down, the lifting of the cordon was delayed as a gas cylinder had to be found inside the building. A robot was used to locate the cylinder but a gas leak in the main Transco pipe was then discovered. Residents from Sir Christopher Court, St John's Street and New Road had to move to B&Bs, Hythe and Dibden Parish Council evacuated The Grove, and Hythe Market was cancelled. In all, the cordon was in place for three days. On 5 December, twenty firefighters were called to a fire in North Road on Marchwood Industrial Estate at 3.30 a.m. The fire was under control by 5 a.m. using jets and hose reels, and it was prevented from spreading to the paper warehouse owned by Premier Shredding. However, the fire ruined two temporary office buildings and two cars, including the business owner's £30,000 Maserati. Fire engines from Hardley, Hythe, and Totton attended along with the water carrier from Lyndhurst.

The *Echo* reported good news on 1 January 2008 as the number of fires in Fawley Inclosure and Forest Front had dropped by 78 per cent during 2007. It was also mentioned that there were 60,000 people living on the Waterside. Hardley attended two house fires in April. The wife and nephew of a 43-year-old man saved him from a burning bedroom when a smoke alarm went off at his house in Heather Road, Blackfield. Also, a ground-floor flat caught fire in Laburnam Crescent in Netley View; three crews fought it for about an hour. At 1 a.m. on Saturday 7 June, two people died and one was seriously injured in a car that collided with a tree between Lepe and Langley. On 10 June a man in his forties had to be airlifted to hospital in Salisbury suffering burns to his upper body after a fire on the schooner *Malcolm Miller*, a vessel he was refurbishing at Hythe Marine Park. Crews from Hardley, Hythe and Beaulieu soon put out the fire, which started at 1.30 p.m. at the former RAF Hythe site. Another serious injury happened around 2 p.m. on 9 July at Normandy Way, Marchwood, where the new power station was being built. A man who had been crushed under a girder was released using equipment from St Mary's Fire Station including a 56-tonne lifting airbag and a hand-operated winch. The £350 million power station was due to open the following year. Hardley firefighters had been called to the site along with Hythe and Beaulieu at 10.50 a.m. on 9 June for a small fire in a boiler house. On 10 July the alarm was raised at 3 a.m. for a fire in the administration building at Noadswood School in Dibden Purlieu. Two crews attended the fire and one hosereel was used, together with two BA sets. A sniffer dog called Saxon was brought in to hunt down the arsonists. There were two arson attacks within forty-five minutes of each other in the early hours of the morning on 18 December, one at an empty semi-detached house at Furzey Close, Fawley and the other at Blackfield Health Centre in Hampton Lane. Two emergencies happened on Charleston Industrial Estate towards the end of the year. Firefighters from Hardley, Hythe, Beaulieu, Totton, and St Mary's dealt with a fire in a waste disposal unit on 23 November and an explosion at Pyros Environmental on 5 December.

The threat from arson had not disappeared, and 2009 was a bad year for this. Three fires were lit on Cadland Estate on 22 September, the worst of which was near a pylon, and this caused the National Grid supply to trip out. Twenty-two firefighters worked with hose reels and beaters, using eight BAS, and they tackled a fire covering 800m by 50m. HF&RS became aware of the incident at 11.30 p.m. and crews from Hardley, Hythe, Lyndhurst and Eastleigh were on the scene until 3 a.m.. Chris Yandell's report read, 'People living in Blackfield and Fawley have told how the late-night incident caused a bright orange glow that lit up their bedrooms. A police spokesman said officers were anxious to hear about anyone who returned home smelling heavily of smoke'. A *New Forest Post* report from 1 October said Steve Foye, the New Forest Group Manager for the Fire & Rescue Service, was taking the incident seriously; also that partnership working had decreased an arson trend in the area before these attacks. However, there were further deliberate forest fires around the beginning of October.

When a dustcart caught fire on the way to Marchwood Incinerator late in October, the two operators managed to jump out just in time. Ministry of Defence firefighters along with crews from Totton and Hardley raced to the incident in Normandy Way, where flames were beginning to set alight the resident's hedges. A Calor lorry caught fire at the Calor depot in Cadland Road just after 9 a.m. on Tuesday 24 November 2009.

It was in a 1-tonne tank half-full of explosive liquid propane gas. Twenty-one firefighters from Hardley, Hythe, Beaulieu, Totton, Redbridge, and St Mary's went to the scene, and staff were evacuated behind a 100m exclusion zone. The *Lymington Times* wrote, 'The accidental fire was caused by a flashback during the cleaning process of an adjacent tank when a tarpaulin caught alight. It was extinguished using a jet.' Two ground monitors were placed on either side of the tank. Thermal imaging cameras were used to check that the tank's temperature was dropping. Incident Commander Dan Tasker commended the crews for their work in the face of 'challenging conditions'. Hardley firefighters were called to Buckler's Hard on 18 December when a barn containing thirty half-tonne bales of hay caught alight at Clobb Farm. In total, sixty firefighters attended, and their duties included shepherding ninety cattle to safety! The incident was in progress for three hours and involved crews from the New Forest, Southampton, and Portsmouth.

An incident happened at Fawley Refinery on Friday 12 February 2010. At 3.15 p.m. a fire broke out in a high-pressure oil feed drum, sending smoke over Holbury for fifteen minutes. An *Echo* report said Hardley, Hythe, Burley and Eastleigh dealt with the fire alongside the refinery's own fire team. There was a large police presence at the incident and several fire officers' cars attended, as did the Command Support Unit and the ICU. Cadland Road was closed until just before 5 p.m., by which time the fire was under control.

The Ford Transit Command Support Unit HX05 JUH is often included in a fire brigade attendance to the Fawley petrochemical industries. Known as 'Command 2', the vehicle is based at Service Headquarters in Eastleigh and is staffed by retained firefighters. (Matthew Leggott)

Tankers returning to the refinery were directed to park in Hardley Industrial Estate until 5.30 p.m., and fire crews from St Mary's stood by at Hardley Fire Station. On 23 April, Hardley crews helped with fire in grassland and undergrowth at Blackhill Road, West Wellow, off the A36, which was put out soon after 7 p.m. Other New Forest and Southampton stations were involved. Some incidents in May required a large fire brigade attendance. A fire that first engulfed a caravan spread to the owner's bungalow at New Road in Blackfield late one night. Fire engines from Hardley, Hythe, Totton and Beaulieu all attended the incident, which was probably arson, along with the SEU from St Mary's. A call to the Tradbe waste site on Charleston Industrial Estate came soon after 5 p.m. on 13 May. Crews from all four local stations plus Redbridge and St Mary's rushed to a fire in 50 tons of woodchip, sawdust, tar, and animal by-products in a silo at the works. Many of the same vehicles returned the next day for an unrelated incident in which 1,500 litres of inflammable liquid was leaking from a storage tank; this was dealt with using foam. On 5 July at 3.30 a.m. a fire erupted in a semi-detached house in Eastcot Close, Holbury. The occupant survived thanks to a smoke alarm. Firefighters stopped the flames engulfing the house next door, where the family had escaped because their cat had woken them. Hardley dealt with a fire in 1,000 bales of hay at Lime Kiln Lane, Holbury on 17 August, a flooding incident at Emsworth using the HVP on 9 November, and an arson attack on an elderly couple's home in Mopley early in December. Unusually, on 28 December the driver of the Hythe Pier train spotted a fire on a 25ft unoccupied fishing boat drifting 50m off the pier. Firefighters from Hardley and Hythe fought the fire, in the boat's engine compartment, from a Southampton harbour master's vessel.

 Of several property fires in the early part of 2011, one at Marchwood on 18 January was the most serious, requiring a six-pump attendance and taking three-and-a-half-hours to bring under control. The HVP unit operating from Hardley went to the Swinley Forest fire in Berkshire early in May. This incident affected 740 acres of forest, 55 per cent of which was damaged by fire or clearing operations, and it required a fire brigade attendance from 2 to 9 May. Arson was still a problem locally, as a report by Chris Yandell on 17 May said fourteen fires had been started deliberately in Hythe, Marchwood and Blackfield in the last four weeks, including three on the same day. In June 2011, Hardley's Land Rover was sent to Standing Hat in the Pignall Inclosure, a six-pump fire involving forty firefighters and requiring the ICU. July was a difficult month, not least for the fire brigade. First a co-responder car hit the Peugeot 306 of a 72-year old woman in Fawley Road whilst driving towards Hythe on blue lights. Then, on 21 July a contract worker at Fawley Refinery was killed when pipes fell off a lorry and crushed him. Hardley Fire Station was called, and there was a significant response from South Central Ambulance Service, including their ICU. An ambulance carrying the man left under police escort but tragically he was deceased on arrival at hospital. A fire in 90,000 tonnes of rubbish at the Blue Haze Landfill Site broke out at 3.30 a.m. on 4 September. Thirty firefighters from Ringwood, Fordingbridge and Hardley attended, and the HVP from Hardley was deployed. There had also been a fire at the site in May. On 25 July an explosion and fire was reported just after 7 a.m. in the laboratory at Marchwood Scientific Services in Marchwood Industrial Park, off Normandy Way. A rapid response to the incident came

Hythe's vehicle, currently the ex-Winchester Volvo FL/Emergency One appliance HX59 AVB, is the second engine to attend incidents within the Hardley area. Hythe's previous appliance, W886 WRV, was named 'Centurion' as it was the 100th Volvo fire engine purchased in Hampshire! (Roger Hansford)

from Hardley, Hythe, Totton, Redbridge, St Mary's and Lyndhurst; the author, by chance, was travelling up the A326 at the time and saw many of their vehicles being deployed.

At 6.30 a.m. on 20 June 2012, a horse owner discovered her mare and newborn foal trapped in mud by the estuary wall at Ashlett Creek. Specialist animal rescue teams arrived at 7.30 a.m. and initially removed the foal, knowing its mother would follow it! In a more serious incident on 12 September, a 44-year-old worker from Hythe suffered crush injuries at the Tradbe site on Charleston Industrial Estate. A hydraulic platform had pushed him against an overhead gantry, and he was airlifted to hospital with life-threatening injuries. Hardley and Beaulieu crews attended along with police and ambulance staff. There was another serious incident at the same site at 8.30 a.m. on Friday 28 September when a 14kg pile of sodium reacted with water. Hardley raced to the scene, followed by Beaulieu, St Mary's, and Totton, but the fire was out before they got there and they left the site just after 10 a.m. Police and a medical Hazardous Area Response Team (HART) were also called, along with a fire brigade Detection, Identification and Monitoring (DIM) unit. On 9 October Hardley and Lyndhurst firefighters rescued two people from a car stranded by floodwater in Balmer Lawn Road, Brockenhurst. At lunchtime on 1 February 2013 fire crews rushed to Long Lane in Holbury where a car and pick-up truck had collided. The car driver, who was cut from the wreckage, complained of back pain and was airlifted to hospital by the Hampshire & Isle of Wight Air Ambulance. On 27 February Hardley's crew attended a serious fire in a farmhouse roof at Hill Top, New Forest. Firefighters from

St Mary's, Eastleigh, Hythe, Beaulieu and Brockenhurst stations also attended. They had to work at high speed to contain the fire, but saved the occupants and a dozen horses from the stable next door and extinguished the blaze two hours later. Another property fire occurred on Good Friday, 29 March, at a thatched cottage in Claypits Lane, Dibden, destroying all of the roof and first floor. Hardley attended along with nine other pumps, three specials, and the ICU, and the crews were on scene for five hours. In the early months of 2014, Hardley's crews rescued victims from two road traffic collisions and a house fire, and dealt with blazes in Blackfield medical centre and on a yacht on Hythe Marina. The varied range of calls attended shows how the risk profile of the Fawley area has developed and how the front-line fire station is vital in this community.

PETROCHEMICALS AND AMMUNITION: FAWLEY'S INDUSTRIAL FIRE BRIGADES AND MARCHWOOD DEFENCE FIRE STATION

The Fawley area has an unusually high number of industrial fire brigades, and they benefit from successful working relationships with each other and with the local authority. HF&RS covers specialised industrial risks, but each organisation also has its own provision, whether in terms of fire crews, fire appliances, or firefighting systems fixed within a plant. A large-scale exercise held in Cadland Road, Hardley, in 2012 showed how local authority fire brigades would tackle a major emergency at Fawley Refinery and its adjoining industries, and is an example of the many exercises which have taken place in the area. Of the people contributing to this chapter, it was a special privilege to meet one who has achieved royal commendation and worldwide status in the industrial fire safety industry. Many of the area's firefighters have worked for more than one of the fire brigades, and the links among them run back into previous generations of serving firefighters' families. Charlie Holmes was Officer in Charge of Hythe Fire Station and the fire team at ISR, while – more recently – Terri Smith ran Hardley Fire Station and the Esso Fire & Response Group. My closing case study of the military fire brigade at Marchwood's Military Port (Sea Mounting Centre) reveals interesting organisational differences compared with county fire brigades, as well as a different job day-to-day. The message is that high risks can be reduced with inter-agency co-operation and effective pre-planning.

Esso Fire & Safety Department, Fawley Refinery

One of the notable firefighters on the Waterside was Barry Browning, MBE, the Chief Fire Officer at Fawley Refinery from 1983 to 1995. He rose through the ranks at Fawley and achieved thirty-nine years of service before he retired. His recollections provide an expert view of firefighting provision and operations at the refinery and chemical plant.

Personnel from Esso Refinery in the early days. From left to right: back row – Charlie Holmes, Charlie Burnett, Colin Falconer, Ray Axell, Bill White, Arthur Orman, Arthur Kitcher, Mike Mullens, Paddy Brown, Dr Ward Gardner, Doug Lindsey, Jack Rayment, Reg Holley, Burt Wheeler, Horace Bonner. Front row – Reg Young, Bob Heaverman, Ben Hendry. In the background are three original 'limousine-style' fire tenders and an AEC Mammoth Major foam tanker. (Barry Browning)

Fawley Refinery Fire Station, early 1960s, showing (left to right): two Dennis F2/Pyrene fire tenders, Dennis F2/Foamite foam tanker, Dennis F6/Foamite water carrier, Commer QX/Pyrene emergency tender with carbon dioxide, Land Rover, and Fordson Thames Control Unit. (Dick Lindsay)

Since retirement from Esso he has published journal articles on fire safety, and lectured on the subject worldwide whilst running his own fire safety consulting firm. He is part of British Petroleum's International Emergency Call-Out Response Team. Barry was awarded the Annual Industrial Fire World Award in the USA for being an Outstanding Professional in 1992, and he received his MBE from the Queen in 1994 for his contribution to industrial fire safety. Barry's predecessor for the 1992 award was Red Adair – famous for fighting the Piper Alpha oil rig blaze in 1988, and the oil well fires in the aftermath of the Gulf War in 1991. Barry had first-hand contact with serious incidents at Fawley Refinery, and describes running for his life whilst tackling a fire in an exchanger that exploded at close proximity. He remembers facing incidents involving acrid smoke, chemical leaks, and fires in plants that were difficult to access due to their height. On one occasion burning oil shot 2,000ft up into the air.

Barry joined the Esso Fire Department in 1957, before there was a chemicals site at Fawley. Originally there were five firemen on each shift, including one station officer, one Sub-Officer, and one leading fireman, so that cover was provided twenty-four hours a day. Volunteers among the shift maintenance workers would back-up the full-time crews, although this practice was curtailed in 1983 according to an *Echo* report.

One of the early Esso Fire Department competition teams with their trophy! Left to right: back row – Basil Burton, Barry Browning, Jim Chiverton, Doug Webb, Roy Axell. Front row – Tony Robbins, Chief Fire Officer Billy George, Paddy Brown. Behind them is an original Pyrene fire tender. (Barry Browning)

Esso firemen practising a trailer pump drill for a Hampshire Fire Service competition. Esso Fire Brigade held the record for this annual competition! Left to right: Billy George (Fawley's first Fire Chief), Doug Webb, Ray Axell, Jim Chiverton, Basil Burton, Paddy Brown, Barry Browning. (Barry Browning)

There were twelve auxiliaries per shift. In the early days the firemen wore oilskins instead of fire tunics, later acquiring Hendry helmets with visors. They trained on open ground on the site of the old AGWI Refinery, discharging fire extinguishers to put out crude oil or naphtha fires lit in half-cut 36-gallon oil barrels. Barry's team was responsible for the site's water mains, which had been laid underground in order to survive blast damage from an explosion. Firefighting in the refinery depends on a high-pressure water supply with sufficient volume to maintain pumping for lengthy periods. Seawater for firefighting and process cooling is drawn through pipes of 60in in diameter on the marine terminal, powered by five pumps each capable of 31,000gpm capacity, and fed into storage tanks holding a total volume of 16 million litres.

Barry remembers some of the equipment used to fight fires in the early days, before large monitors were mounted on the top of fire engines. There were three foam production units made by either Foamite or Pyrene, and two had 36hp Rolls-Royce engines with open cabs – so the drivers got wet in bad weather. Barry said there were not enough monitors at first, and those they had were 'awful beasts' because they had to be assembled by hand. There were four portable foam towers kept on a trailer. The crew had to tow the trailer up to a storage tank fire, lift one of the towers over the tank wall, and secure it in place before winching the tower up and over the side of

This early Fawley Refinery fire appliance photographed against the installation it protected was a Dennis F2 pump/foam tanker built by Foamite in January 1950. It was registered PYO 328. It was still on site as late as 1970. Note the hoses, branches, and monitor trailer it carried. (Dick Lindsay)

This Esso Fire Department Fordson 1952 Control Unit began life as a Civil Defence Corps Signal Office. During a major incident it would be crewed by Hampshire firemen from Lyndhurst, who would lay out waymarkers so other vehicles could follow them through the refinery to the incident scene. (Ken Reid)

the burning tank. The foam production came from 500gpm Coventry Climax trailer pumps, of which there were four. These could be connected to the hydrant, and chemical foam was produced when acid and alkaline powders were added to chambers on each respective side of the pump. Barry describes how Fawley produced the original trailer-mounted monitor.

It was towed behind a Land Rover and based on equipment initially fitted to firefighting tugs like *Culver* and *Atherfield* – these were moored at the marine terminal in readiness, and later replaced by *Gatcombe II* and *Vecta II*. The large trailer monitor, known colloquially as the 'Prontosaurus', was capable of producing either a jet or fog spray, and it was later commercially available from manufacturers like Chubb and Strebor who worked to Esso's specifications.

Three generations of Esso Fire Chiefs: Barry Browning (rear) replaced Colin Falconer (front left), who replaced George Nash (front right). (Barry Browning)

Barry was Fawley Refinery's last Fire Chief, following Billy George, George Nash, and Colin Falconer. Barry has a firm belief in professionalism. He knew he needed to introduce a new uniform and improve his crew's morale. He replaced the Seibegorman one-hour oxygen sets with a new brand. He extended the fire training ground and added fire simulators. His requests for new firefighting equipment or water mains were some-times turned down by Refinery bosses, who needed costings to be strictly competitive. He would always say that Fawley's most recent tank fire was tomorrow! Barry often decided to respond to fires at other refineries when he heard about them on television. He went to Milford Haven in 1983 where a crude-oil storage tank 75m in diameter (capacity 95.45 mil-lion litres) was on fire, and he gave a presentation on the incident when he returned to work at Fawley. This lead to improvements in Fawley's provision for storage tank fires, including the purchase of new fire tenders with greater pumping capacity and larger portable moni-tors, a greater on-site stock of foam concentrate and an improved water supply in tank farm areas. In this way he improved cover locally, and enhanced his international reputation in the fire safety industry. Each incident, wherever it occurred, could offer learning points that would make the industry safer for the future. Barry's excellent reputation spread further after his advice helped control a multi-tank fire in Greece in 1986. One of Barry's publications (March 1997) sets out his method for petrochemical firefighting:

It is universally recognised that to effectively control hydrocarbon fires you need MASSIVE COOLING in the shortest possible time, with a minimum of manpower. Without a fast response, coupled with the use of suitably-designed fixed and portable firefighting equipment handled by well-trained personnel, minor incidents may quickly turn into major incidents.

In the early 1970s, a pair of Ford D1000 fire tenders was the first to replace the early generation of 'limousine-style' appliances at Fawley Refinery. This 1972 example, registered GHO 149K, was built by HCB-Angus, but the 1970 vehicle had bodywork by Merryweather. (Roger Mardon)

Fawley Refinery identified the need to transport portable monitors after a severe fire in a power former in 1969. YOT 31J was a 1971 Bedford TK vehicle with a Ratcliff tail lift that could carry monitors to the site of a fire. The 1980 Bedford TK registered XTP 531V also performed this function. (Roger Mardon)

Barry describes how Fawley's fleet of firefighting vehicles improved in response to learning points from particular incidents in the petrochemical industry. For example, a Bedford TK flatbed lorry carrying portable monitors was implemented after the severe power former fire at Fawley in 1969. At the height of this fire, a volume of 48,000lpm was required. Barry secured financial backing to replace the foam and water fire tenders when he took over as Fire Chief in 1983. The new tenders shared their specification with fire vehicles at an Antwerp Refinery; each had an 8,000lpm pump, and a roof monitor with a reach of 70m. They were initially made by Carmichael, and then by HCB-Angus, with the cabs and driveline by Renault-Dodge and then Mercedes-Benz. In response to the Milford Haven fire, Barry acquired a towable 24,000lpm diesel pump and two hose trailers to supply large-capacity foam monitors. With the advent of high-volume monitors, firemen could work in safer conditions at a greater distance from a burning tank. Refinery fireman Denny Blackmore designed a new badge for the Fire Department, consisting of an Esso logo ringed by a fireman's belt and surrounded by a crest. The vehicle fleet included a Ford Transit general-purpose van and a Fire Chief's Mini, both with the red livery, badge and blue flashing lights. Barry acquired a Foden articulated tanker that was being retired from Esso delivery duties at the Avonmouth Depot. An Esso fireman collected it and it was painted in fire department colours for use as a foam tanker at Fawley. Barry remembers having it hidden behind the bitumen plant and then parked outside the refinery manager's office; he told the manager 'it's yours'!

AEC Mammoth Major foam tanker MXD 69 of the Esso Refinery Fire Department, built in 1952. As with the articulated foam tanker later based at Fawley, it began life as an Esso petrol tanker. Firemen Barry Browning and Gerry Holloway drove the tanker to the Chubb works at Luton to refill it with foam required for the *Pacific Glory* disaster in the Solent in October 1970. (Barry Browning)

Multi–Agency Firefighting at Fawley Refinery

Today, the Esso Fire & Response Group is run by Chief Fire Officer Steve Tolley, who replaced Terri Smith – a former retained Sub-Officer at Hardley Fire Station. Esso's equipment has been modernised, including the purchase in 1999 of two 6,000gpm foam cannons known as 'six guns', and a sophisticated Mercedes-Benz fire tender new in 2009. Throughout its history, Esso's fire brigade has dealt with small fires in the petrochemical complex where the attendance of county fire engines was not required. But support from the county can be sought if needed; liaison between the two brigades is generally good. HF&RS PDA to the refinery and chemical plant in 2012 was just two fire engines. However, this can be 'made-up' to a larger number if necessary; the automatic response was bigger in the past, involving up to ten appliances initially. Chas McGill at Hardley Fire Station can remember an incident when over forty fire engines were parked on the refinery's 'C Avenue' outside the Esso Fire Station, the crews waiting in safety there before being deployed. One reason for a significant attendance of Hampshire vehicles would be to increase manpower at an incident, as the first wave of firefighters would quickly become tired fighting an intensive petrochemical fire. Another reason would be to boost water pressures for sustained pumping operations. There are 6-in faucets, known as 'storz connections', where Hampshire fire engines can link with the refinery's water main. Above them, a sign reads: 'Cooling water for emergency fire fighting use only, keep access clear'. A tally system for the refinery allocates blue tags to water tenders and red tags to water tender ladders. Water tenders are generally more useful inside the petrochemical plant, the water tender ladders in the community outside. Ideally, the Hampshire fire engines stationed close to Esso are fitted with a 1,000gpm pump, suitable to feed two 500gpm monitors for petrochemical firefighting using a 'Y' connection splitter. There have been many successful training exercises when both parties have worked together, including a command, control, and communications exercise held in Esso in 2009.

Exercise Shannon, May 2012

The most recent major exercise in the area was 'Exercise Shannon', held on Friday 4 and Saturday 5 May 2012. Hampshire Fire & Rescue Service organises an annual county exercise, and this one dealt with the management of petrochemical risks at Fawley. The exercise was designed to meet training needs and improve management issues from previous incidents and exercises. The aims of Exercise Shannon, organised by Watch Manager Chas McGill, and drawing on his experience as South East regional co-ordinator for HVPs, were as follows:

- To provide a water supply to multiple top-tier COMAH sites on the Waterside area in Hampshire.
- To test airwave multi-agency interoperability between Hampshire Constabulary and other agencies.

Arriving at Fawley Refinery in 2010, this new fire tender HX59 FHE was a Mercedes-Benz 1429 Atego adapted by Protec-Fire. (Matthew Leggott)

- To augment existing water supplies in Esso Fawley using HVPs, and to test HVP output at Hythe Tanker Terminal.
- To test a full convoy of HVPs and support vehicles.
- To test HF&RS response to a road traffic collision involving a tanker leaking LPG.
- To establish Silver Command at Hardley Fire Station.

It was also important to establish a functioning asset-holding area, so that vehicles were not parked at Hardley Fire Station or near the refinery entrance in Cadland Road. A huge number of personnel were involved from many different organisations, and all the simulations were risk-assessed and carried out in conjunction with Esso's safety rules. I accompanied personnel from specialist fire journals to a briefing on the morning of Friday 4 May, and watched the exercise scenario unfold on the Saturday.

The initial call was made to Hampshire Fire Control from Esso Fawley at 1 p.m. on 4 May, saying their saltwater pumps on the marine terminal had failed. This compromised their ability to provide water for their own fire main or for those of other nearby COMAH sites. They had also lost the supply of cooling water for their process units and therefore initiated a Refinery shut-down. Esso notified the other COMAH sites of the situation. Hampshire Fire Control informed the 'on duty'. Waterside Station Manager and the HVP advisor to attend an emergency briefing at the refinery, along with a representative from Hampshire Police. The personnel were Station Officer Mick Thompson, Watch Manager Chas McGill and Inspector Mike Batten, who arrived at Refinery Gate 1 and were escorted to the meeting with Esso bosses. Their decision was to deploy HVPs to construct a temporary fire main. The HVP team, comprising firefighters from Hardley, Hythe and Ringwood fire stations, was called to utilise the HVP resources kept at Hardley. A call was also put out for the Dorset HVP team to attend. The crews were on the Esso site between 4 and 6.45 p.m. on 4 May, with the task of establishing a supply of fresh water drawing from a 3-million-litre lake known as the 'catchment area' at the bottom of the refinery.

Command teams were put in place at HF&RS Headquarters and in Esso, and by 11 p.m. on Friday it was decided that additional HVPs should be requested from the National Fire Control Centre as a precautionary measure. The mobilisation of HVPs on a national scale involved units from Royal Berkshire, Surrey, Oxford, and the Isle of Wight assembling at Hampshire Fire Headquarters in Eastleigh, where overnight accommodation and feeding were provided for the crews. The most efficient way for the resources to reach the refinery was to travel as two separate blue-light convoys escorted by Hampshire's Roads Policing Unit on the morning of 5 May. Five police motorcyclists cleared a path for the convoy by riding ahead and stopping the traffic at each junction. Journalists were allowed to travel in the cabs of the HVPs and cameras were placed on some of the vehicles, including the police motorcycles. The line-up of each convoy, detailed as follows, was dictated by the order in which vehicles would be needed once they reached the Tactical Holding Area (THA) – next to the Calor Gas site in Cadland Road.

Oxfordshire Fire & Rescue Service provided the leading vehicles for the second convoy that travelled from Eastleigh HQ to Fawley Refinery on 5 May 2012. (Roger Hansford)

BMW motorcycles from Hampshire Constabulary's Roads Policing Unit escorted the Exercise Shannon convoys from Eastleigh to Fawley. (Roger Hansford)

Convoy A – leaving
8.30 a.m., 5 May 2012
Dorset Water Tender
Dorset Officer's Car
Surrey HVP
Surrey Hose-Layer
Surrey Support Vehicle
Surrey Officer's Car
Isle of Wight HVP
Isle of Wight Support Vehicle
Babcock Mobile Maintenance Unit
HART Incident Command Unit
HART Land Rover Discovery

Convoy B – leaving
9.30 a.m., 5 May 2012
Oxfordshire Water Tender
Oxfordshire Hose-Layer
Oxfordshire HVP
Oxfordshire Support Vehicle
Oxfordshire Officer's Car
Royal Berkshire HVP
Royal Berkshire Hose-Layer
Royal Berkshire Support Vehicle
HART Land Rover Discovery

Once the HVP resources were at the scene, their task was to extend the overland water main initiated the day before by the Hampshire and Dorset teams. The Surrey and Berkshire HVP teams brought the hose-line up Cadland Road so the Oxfordshire crew could deliver water at Hythe Tanker Terminal. Chas McGill designed this part of the exercise to improve on a previous simulation he took part in at the Avonmouth Esso Terminal, and he reached his objectives. The Berkshire team required escort by a police BMW X5 whilst laying the hose line against the flow of traffic to and from the other COMAH sites. Once in place, the relief water main was pumping to a height of almost 30m above

A Royal Berkshire HVP crew at work in Cadland Road, Hardley, collecting up the high-volume hose line as two police patrol cars leave the simulated accident scene at Exercise Shannon. (Roger Hansford)

sea level. As this was proven to work with one hose-line, the tactic could theoretically be duplicated for three or four lines, and the pumping maintained for an indeterminate extended period. Even with the exhaustion of the catchment area, a supply could be sourced from Southampton Water.

Whilst the crucial pumping aspect of the exercise continued, Hampshire Fire Control received emergency calls to other simulated incidents in the vicinity. In both cases, conventional fire appliances could deliver water from the fire main put in place by the HVPs.

With the bridging fitted, the high-volume hose line crosses the road. The emergency services have arrived but remain outside the exclusion zone set up around the 'leaking' tanker on 5 May 2012. (Roger Hansford)

The function of this monitor is to disperse the supposed gas cloud from the staged tanker crash at Exercise Shannon, using water from the high volume hose line. (Roger Hansford)

At 9.30 a.m., Esso made an emergency call reporting smoke issuing from chemical containers at the Nalco site within the refinery boundary. At 9.33 a.m. the appropriate crews were briefed, and appliances deployed to Refinery Gate 1 under escort from Exxon-Mobil security personnel. On arrival at the site, HF&RS crews discovered that palletised tanks containing chemicals were undergoing an exothermic reaction. Firefighters from Beaulieu, Fordingbridge and Burley worked with Esso Fire & Response Group to bring the situation under control, drawing from the relief water main, which at this point was extended into the refinery by the Isle of Wight HVP team. This part of the exercise was useful in training the crews for an incident inside the refinery complex, particularly in terms of the multi-agency liaison required at each stage.

At 10.30 a.m., just as the situation at Nalco was about to be declared safe, an emergency call was received from the Flogas depot, reporting an explosion and gas leak in Cadland Road. All agencies were informed, and three pumping appliances were despatched from the THA. Personnel from HF&RS, Hampshire Constabulary, and a medical Hazardous Area Response Team (HART) were on the scene soon afterwards, as a police helicopter circled overhead. What faced the crews was an overturned tanker leaking LPG that had toppled when turning into the Flogas depot, creating three 'casualties', including one amputee. Officers commanding the incident for all three emergency services decided a strategy by 10.45 a.m. The casualties were attended to and removed from the road by a team wearing breathing apparatus, comprising four firefighters and two members from HART.

Three casualties and a gas leak: the mock incident involving a tanker at the entrance to the Flogas premises in Cadland Road on 5 May 2012, seen before emergency crews arrive. (Roger Hansford)

Two contrasting 4x4s and fire brigade hoses at the entrance to the Polymeri Europa premises during Exercise Shannon. On the left is a BMW X5 police car and on the right a Land Rover Discovery vehicle of the HART operated by South Central Ambulance Service. (Roger Hansford)

All other emergency personnel were kept at a safe distance behind a police cordon, where the casualties received treatment. At 11.05 a.m. the Hazmat Officers decided to place a water monitor on the stricken tanker to disperse the gas cloud, again drawing from the relief water main. Finally, the Police Forensic Investigation Team moved in to gather evidence, during which they put new investigation technology to the test. Although this part of the exercise happened in close proximity to the Flogas Depot, it simulated an incident that might have occurred on any road in the county. The aims of Exercise Shannon were all met, and it was comforting to know the fire brigade could rise to deal with a serious petrochemical incident, whether accidental or caused by terrorism.

BA crews from St Mary's Fire Station in Southampton prepare to enter the exclusion zone at Exercise Shannon to deal with the emergency. (Roger Hansford)

Firefighters approach the stricken tanker, pulling both first-aid and delivery hoses with them. (Roger Hansford)

The most severe casualty, an amputee, is taken to safety by firefighters and medical staff wearing breathing apparatus. (Roger Hansford)

The Industrial Fire Brigades on Charleston Industrial Estate, Hardley

The county fire and rescue service will respond to the industrial sites on Cadland Road and Charleston Road – adjacent to Fawley Refinery – but some sites have their own arrangements in place as well. Usually a commercial company's firefighting team is made up from several full-time day firemen and an officer on each watch, supplemented by trained plant workers on each shift. Like retained firefighters, the shift workers would switch to firefighting in an emergency.

The former Monsanto Chemicals site was a good example of this duty system, with one whole-time fireman per shift supported by trained shift workers. Later they changed the system and relied on the trained shift workers, co-ordinated by a full-time day firefighter and a safety officer. Now derelict, the site produced polyethylene plastic granules, using ethylene gas from Esso. There was a small fire station with a red roller shutter door on the premises. The fire appliance was a Land Rover 109 registered UOR 92 and converted by Carmichael. A photograph of the vehicle appeared in a 1963 book, *Fire Engines* by Michael Rolfe, where it was described as a 'miniature fire engine' and said to be 'operated by a factory brigade in the South'. An *Echo* report of January 1968 said the site's no. 2 reactor had to be shut down and de-pressurised after a Saturday night fire that was 'tackled by the works' brigade', with Hampshire appliances from Hythe, Lyndhurst, and Beaulieu in attendance to provide cooling water. Similar incidents occurred on 22 September 1972, 9 February 1975, and 23 January 1976, with fires being extinguished by the fire team or the automatic sprinkler system with Hampshire firemen standing by. One member of the Monsanto fire team was the father of Alan Nichol, now a crew manager at Hardley Fire Station.

Another site with its own fire team opened as Union Carbide in 1959, changing its name to BP Chemicals or 'Hythe Chemicals' in 1976. On a 25-acre site, the plant produced solvents and detergents as well as anti-freeze liquid and brake fluid using ethylene piped from the refinery. Since the 1990s it has traded as Inspec UK, Laporte, Degussa, and then Cognis UK. Under the site's present name of Geo Speciality Chemicals it has no fire team. However, fifty of their 150 employees are trained to give a 'first-aid' firefighting response, and they undertake this training at the Warsash Marine Terminal. There is a fixed firefighting system with a ring main around the site, fed by two diesel pumps of 9,600lpm, and there are twenty-six fire hydrants. Many of the site's sprinklers and deluge systems are activated automatically by gas detectors. The company's only emergency vehicle is a large van used for transporting personnel and equipment to remote corners of the site.

One of the largest chemical sites on Charleston Industrial Estate opened as International Synthetic Rubber (ISR) in 1958; its name later changed to Enichem Elastomers and then to Polymeri Europa UK. Rubber products, elastomers and PVC polymers are manufactured on the 54-acre site, which includes the risk of LPG and other chemicals, and requires both

The Hardley chemicals plant which opened as Union Carbide in 1959 traded as Cognis UK from 2003 to 2011. Their site fire appliance was a four-wheel-drive Renault-Dodge/Carmichael tender with red-and-white check and blue striping and both orange and blue lights. The current emergency response vehicle is an LDV Maxus five-seater van with a rear compartment for the fire kit. (Matthew Leggott)

Based at the premises of Polymeri Europa UK Ltd, this 1990 Volvo FL6.14/HCB-Angus appliance G165 UPO was previously stationed at Andover and also Yately in the Hampshire Fire & Rescue Service. (Matthew Leggott)

foam and water for firefighting operations. Today the plant's emergency vehicle is an ex-Hampshire Volvo FL6.14 built by HCB-Angus, but their previous fire engines included a Bedford TK water tender, and before that a Fordson Thames. The latter was pictured in a 1962 press article saying ISR was 'a local company which has paid much attention to providing the maximum protection against fire', including its own fire brigade:

> The high-speed operation of the ISR works fire brigade has been proved not only in constant drills and practices at the plant, but in successful participation in competition with other works fire brigades around the country. The ISR works fire brigade is proud of its very close liaison with the public fire brigade, who are automatically called every time the factory fire alarm is operated.

On the scene in seconds

The part-time retained members are enrolled from all sections of the factory – engineering, production, laboratories, administration – and are all given drills and practises on a regular weekly basis. Should a fire or any other emergency arise, the fire alarm is automatically operated and the part-time firemen immediately come under the direct control of the professional fire officer on duty. These men, with their appliances and equipment, can arrive at the scene of any incident within seconds day or night.

Traditionally, the Chief Fire Officer and many plant workers at the site were retained firemen from Hythe Fire Station. An *Echo* article of 5 October 1962 said Charlie Holmes was retiring aged 55 as station officer at Hythe, but would continue as Chief Fire Officer for ISR. He had become a fireman in Margate in 1933 and had been in charge of Hythe since 1948. He was presented with a barometer during a dinner dance at the Montagu Arms in Beaulieu. Pam Whittington (1998) said Charlie had partly designed the layout for the ISR chemicals site, and had also worked for the British Powerboat Company and Esso Fire Brigades. He had been in charge of the Powerboat Brigade before taking up the post with ISR. Jim Haines, station officer at Hythe 1976–81, and Gordon Willis, Sub-Officer in charge of Hythe 1981–84, were also ISR firemen. Colin Falconer had recruited Gordon into the Hythe crew. Gordon was one of four full-time Sub-Officers at ISR – each one leading a shift – along with Jim Haines, Mick Rouse, and Colin Fenn. Colin Fenn worked at ISR for thirty-five years, and was chairman of the Southern Branch of the British Fire Services Association (BFSA). Today, Phil Gittings is the Sub-Officer in charge of Hythe and the Chief Fire Officer at the former ISR works, now Polymeri.

The ISR Sports and Social Club in Blackfield often served as a venue for firefighting competitions. These were important events in a fireman's calendar, particularly during the 1960s and '70s, and the ISR crews had a good record of success. One article documents the BFSA (Southern Branch) Annual Fire Brigade Competitions held in the snow on 5 April 1975, when Charlie Holmes was in charge. It reported that the competition season

Bedford TK/HCB-Angus works fire tender, ISR Fire Brigade, registration YOR 713J. It was equipped with an 1,800-litre water tank, 100-litre foam tank, and Bayley 105 (10.5m) wooden extension ladder. On the left is Sub-Officer Gordon Willis, with Chief Fire Officer Charlie Holmes on the right. (Trevor Fenn)

Bedford TK/HCB-Angus Works' fire tender, ISR Fire Brigade, with 10X (middle-sized) foam branches in use. Colin Fenn, holding the branch on the right of the photograph, is working with one of the other Sub-Officers to dispense the foam. (Trevor Fenn)

had started with the ISR crews gaining two firsts, two seconds, and the 'best aggregate' shield, and they set a competition record time with seventy-one seconds for the 'four-man with officer' event. There were six separate events, for public and industrial fire brigades, with teams from Glastonbury, Portishead, Berkeley, and Brighton. Chief Fire Officer Clayton, the National Chairman of the Association, had travelled to the event from Grangemouth in Scotland. The article offers a snapshot from the past which speaks of a thriving fire brigade that was always honing its skills with competitions and training.

The training was necessary, as the ISR site saw serious incidents involving dangerous chemicals. A fire in a 500,000-gallon latex tank broke out early on the morning of 13 June 1974, requiring a Hampshire Fire Brigade attendance of eight appliances plus foam tanker. These were on the scene just after 6 a.m. and the fire, which emitted thick smoke, was out two hours later. Foam was pumped inside the tank and cooling water sprayed on the outside. The ISR Works Brigade assisted with the operations, and the Esso Works Brigade provided the foam. A fire in the site's Rubber Drying Unit ignited just after 7 a.m. on 31 January 1976. Five fire crews from Lyndhurst, Hythe, and Beaulieu, together with the works brigade, brought the fire under control in an hour, although one fireman suffered unconsciousness after a head injury and was taken to hospital. A similar, but less severe fire followed on 19 October. A fire in a product filter at ISR late on 16 January 1977 caused £30,000 worth of damage, and required the attendance of forty firemen for two hours. Thirty firemen were called on 28 June 1978 when a fracture occurred to the pipe carrying

ISR Competition team with their trophy, photographed at Fawley Power Station. George McCrae is on the left, with Malcolm Lovett at the rear, Sub-Officer Colin Fenn holding the trophy, and Ted Simpson on the right (dark glasses). ISR continued to use black helmets after Hampshire had switched to yellow. (Trevor Fenn)

butadiene – a raw material used to manufacture rubber – from the refinery into the works. Both ISR and Esso were shut down because of the leaking fumes, and cooling water was sprayed on the pipe. Minor fires occurred at the site in January 1978 and November 1987, both handled by the site's own team with Hampshire crews standing by. A more recent fire took place on 28 November 2006 in an industrial drying oven, bringing twenty-five firefighters to the scene. Many of the incidents occurred before Hardley Fire Station was built, but to this day the site maintains its own fire brigade.

Fawley Power Station Fire Service

The oil-fired power station at Fawley, which closed in March 2013, is an interesting part of the story of firefighting on the Waterside. The power station fire team was disbanded during the mid-1990s, but the site kept fifteen Siebegorman BA sets in case of need, and the plant maintained trained first aiders among its workforce of fifty-five. For many years the power station supported local emergency services by offering them somewhere to train. A rehearsal for dealing with chemical, biological, radiological and nuclear incidents (CBRN) was held at the site on Sunday 29 October 2006, when a convoy of HF&RS vehicles travelled from Fire Headquarters at Eastleigh. The convoy included Hampshire's ICU and Emergency Catering Unit, both made by Mercedes. In other exercises on the site, HF&RS firefighters made a water relay system and dealt with a simulated tank fire, using vehicles from Eastleigh and Southampton. On one occasion, portable pumps from HVP vehicles were airlifted from the power station into Fawley Refinery using a Chinook

ISR Fire Team outside the fire station in the 1970s, with their competition trophies. Chief Fire Officer Charlie Holmes sits at the centre of the front row. In the middle row stand (left to right): George McCrae, Mick Rouse, Gordon Willis, Jim Haines, Colin Fenn, Alan Webb, –?–, Terry Boulton. The first two men on the back row (left to right) are Den York and Dave Mountford. (Trevor Fenn)

helicopter! This was for a tank fire exercise in November 2007 that involved five county fire and rescue services. Hampshire Constabulary and the Royal Marines Special Boat Service also trained on the site. Its open spaces and variety of working heights were among the training assets it offered.

The power station – built to run off fuel oil piped from the Esso Refinery – opened in 1965, and operated under a number of different organisational umbrellas. The Central Electricity Generating Board (CEGB) became National Power, which split when the functions of generation, transmission, and distribution of power were separated. After the split Fawley was run by Npower, under the German group RWE. Shortly before the power station closed I spoke to Paul Freeman, who was in charge of maintenance and safety there. The site included a number of risks such as the presence of chlorine used to clean cooling water, and the transfer of fuel oil from ship or road tanker in later years. By this point Fawley Refinery was unable to supply the power station with residual fuel oil by pipeline as had been intended. The generating turbines and other plant equipment were protected in the first instance by automatic fixed systems to deliver firefighting water. These would only have been used in an extreme situation. There were also banks of fire extinguishers to project carbon dioxide into cabling compartments in case of fires there. The fire water system was fed by two 47,000-gallon water tanks, backed up by eight 100,000-gallon tanks. There were a series of different hydrants delivering either fresh or salt water.

The fire team was set up by Jack Heath, who later moved to Berkeley Nuclear Laboratories. The power station had three full-time firemen, officially known as 'auxiliary plant attendants' (APAs). They were Doug Willett, Phil Farrell, and Phil Melton. Doug Willett's son now serves as a firefighter at Redbridge Fire Station in Southampton. Phil Farrell played a big part in the Fawley Division of St John Ambulance over the years, and returned to the power station on shift work the week after he retired as an APA. In addition to the main team, there were sixty-seven retained firemen drawn from the operations, maintenance, and services sections. They worked staggered shifts across a seven-day period. The maintenance workers wore yellow helmets, the day workers white helmets, and the shift operators grey helmets. The fire team's equipment included up to 300 hoses, thirty-nine sets of BA and many different hose branches. The next nearest CEGB Fire Station was at Didcot, and the company fire officer – based in Bristol – was Frank Williams, a former Chief Fire Officer of Avon Fire Brigade.

The Fawley site's original two-bay fire station was by the main gate – it remains home to an appliance belonging to the Hythe Fire Engine Preservation Society, and it used to house the Fawley Division St John Ambulance vehicle. The main entrance was some

These activation switches were part of the automatic firefighting system designed to deliver water if needed into the main turbine hall at Fawley Power Station. (Roger Hansford)

At Fawley Power Station, this bank of fire extinguishers would automatically have dispensed carbon dioxide onto a fire in electric cabling. In the corridor outside, sensors monitored the gas levels to ensure a safe atmosphere for workers. (Roger Hansford)

The first fire station at Fawley Power Station: next to the main entrance and security gate but some distance from the major fire risks. At the time of writing, a charity fire engine of the Hythe Fire Engine Preservation Society was still kept inside; another of their vehicles had been stored at Esso Hythe Terminal. (Roger Hansford)

distance from the power station, and the fire team members used to enjoy coming down the road on 'blues and twos'. It was not so much fun when they moved to a new fire station early in the 1990s: this was right by an entrance to the basement – where most fires occurred. The new station was better for response times to incidents, but was on two different levels and had less space than the old one.

Among the fire appliances based at Fawley Power Station were a large foam unit, a Whitby Warrior 4x4 that carried fire extinguishers, and two Land Rovers, one registered simply 'FAWLEY 5'. More recently there were two light fire appliances: a Dodge 50 and a Ford Transit. After service, each of these was sold for preservation at a value of around £5,000, but soon found its way onto the forecourt – much to the annoyance of power station employees! The county fire brigade could back-up the power station Fire Service, and there was a 'gentleman's agreement' with Esso – the refinery would provide additional equipment for a major incident.

Several fires broke out when the power station was being built in the 1960s, and there were many fires in turbine bearings during the early days of running. Often the fire team would receive six or seven 'shouts' per day. At this time, the station was generating power continuously and little maintenance was being carried out. Sometimes oil caught fire underneath the boilers, or between boiler cladding and the wall. On one occasion, a member of the fire

Situated adjacent to the turbine hall, the lower (disused) building with red sliding door was Fawley Power Station's second fire station. (Roger Hansford)

team deliberately lit a fire in part of a huge crane used in the turbine hall. Breathing apparatus was required to deal with it. Needless to say, the perpetrator was caught, and admitted he had done it to break up the night shift! On another occasion, six Hampshire fire engines were called to the site for a serious turbine fire. Deputy Chief Fire Officer Harold Stinton was in charge, and he asked a CEGB employee to shut the turbine down. The gentleman refused as he wanted to reach a record for running the turbine continuously over two years. He said the fire could be contained using just the power station fire team. However, Harold Stinton's authority for managing the incident was greater, and he got compliance by threatening to telephone the Home Office if the turbine was not shut off! Once this had taken place, the Hampshire crews could tackle the fire. The incident did not sour relations too much, as Harold was given a dinner at the power station when he retired.

Echo reports on incidents at the power station frequently said the plant's own firemen had controlled it before the arrival of Hampshire crews, the latter brought in as a stand-by measure. This was the case for fires in April 1969, August 1971, March 1972, January 1974, March and November 1976, February 1978, and February 1986. An article dated 17 December 1988 said three fire engines made a 'wasted journey' to the site! For most incidents, the usual cause of fire was a leak of oil close to one of the turbines and they were attended by five or six Hampshire engines. A boiler fire on 10 November 1979 was extinguished by excluding oxygen supplies, and a fire in a generator unit transformer on 13 April 1981 was put out by the fixed sprinkler system, with Hampshire crews standing

Foamite's unusual conversion of a Bedford TK can be seen in this view of Fawley Power Station's foam tender, WAE 610H. Note the two-tone sirens and search-light. The vehicle was new in 1965 but, along with the Land Rover pump WAE 611H, was not registered until 1970. (Roger Mardon)

by on both occasions. Smaller fires could be allowed to burn themselves out, but this was not so for larger ones. Seven county fire crews under the command of Assistant Chief Fire Officer Leslie Cummins 'fought a two-hour fire with foam and water' at the site in May 1974, after hydrogen leaked from a transformer and ignited in the housing of no. 2 generator. Hampshire crews worked alongside the power station team on 4 October 1979 when an outbuilding caught fire. Started accidentally by a blowtorch during demolition of the chlorine-cleansing area, the fire was under control in fifteen minutes.

The Fawley Power Station Fire Service regularly took part in firefighting competitions run by the British Fire Services Association (BFSA) and the Industrial Fire Protection Association (IFPA). The IFPA final was held at the London Fire Brigade Headquarters in Lambeth. Fawley often won the trophy for the best turned-out team. Their main competitors came from Thames Valley Police, Avon Fire Brigade, and Rawldon Colliery near Ashby-de-la-Zouche. Fawley were supposed to return the trophy a fortnight ahead, but always walked into the building holding it on the day, giving them a psychological advantage over the other teams! They used another trick to win the fire extinguisher drill – usually set up to involve a burning liquid. With traditional foam, firemen directed it above the fire so it would trickle down and eliminate oxygen to quell the flames. In contrast, the Fawley team used Aqueous Film-Forming Foam (AFFF), but without telling anyone. They gained a considerable time advantage as AFFF could be sprayed directly onto the fire. Nobody realised why the Fawley team kept winning!

A Dodge 50 fire appliance of Fawley Power Station Fire Service, built in 1980 and photographed whilst in preservation in 2010. (Roger Hansford)

Marchwood MOD Sea Mounting Centre

A military base was established at Marchwood as a point of departure for the D-Day landing operations in 1943, and for the Falklands War in 1982. The site is now known as the Ministry of Defence Sea Mounting Centre, formerly Marchwood Military Port. It is the UK's only military port, and comprises 289 acres of land with three main jetties, named 'Falkland', 'Mulberry', and 'Gunwharf'. A base for 600 troops and 150 civilians, it is run by 17 Port & Maritime Regiment Royal Logistics Corps. The main purpose of the site is to transfer explosives from land to sea; they are carried abroad to warzones on the ships *Anvil Point*, *Hurst Point*, and *Hartland Point*. This operation has to be covered by an on-site firefighting crew in case of accidental detonation. An Army Fire Service station opened in temporary accommodation at the Port in 1975, replaced by a permanent building in 1988.

Before Marchwood had its own fire station, crews from elsewhere were put on stand-by when explosives were being loaded there. For example, Bill Farr used to come from the Army Fire Service station at Bordon as part of a crew of three. He remembers several other firemen he knew: Station Officers Cedrick Cousins, Colin Edgar, and Arthur Brooker, Sub-Officer Derek Shinn, and Leading Fireman Alec Donald. Off-duty Hampshire crews using a reserve appliance used to stand-by at Marchwood; this was organised from Lyndhurst, but involved Hythe and Totton firemen as well. After Southampton became part of Hampshire Fire Brigade, the cover work was also offered to crews there as a paid overtime opportunity funded by the military. By 1974 the army needed to open a fire station on the site. They erected a portakabin and temporary appliance hut in April 1975,

Defence Fire Services winning Pump Competition Team at the Royal Military College, Shrivenham, 1990. Left to right: back row – Leading Fireman Martin Head, Fireman Arthur Parrot, Fireman Trevor Fenn. Front row – Fireman Paul Watton, Fireman Mark Sillett. (Trevor Fenn)

and stationed their own firemen there. The county fire brigade could still be called in to provide back-up during an emergency.

Derek Turner was a fireman at Marchwood from 1977. He had given up his career as a sales rep and joined the War Department Fire Brigade at Donnington in 1961. He was made up to Leading Fireman in 1968, and to Sub-Officer at Marchwood in 1977. His father had also been a fireman. Derek served at Marchwood until 2002, first running Blue Watch and then White Watch. He enjoyed his forty-one years in the Fire Service, and found Marchwood an easier place to work than Donnington, where there had been some serious fires. He also took up the opportunity to work abroad, serving in

Temporary accommodation: the first Defence Fire Station at Marchwood was housed in a temporary building introduced in April 1975. (Trevor Fenn)

The Carmichael appliance 67 HF 99 was housed in temporary garaging facilities at Marchwood Sea Mounting Centre before the permanent fire station was built in 1988. The chimneys of the former Marchwood Power Station are in the background of the photograph. (Trevor Fenn)

Northern Ireland in 1971, Bosnia in 1999, and Kosovo in 2001, including six months as Station Commander. He said the job had good benefits – such as moving allowances – and you could be successful if you kept fit and 'kept your nose clean'. He enjoyed being in charge of the team of firemen. However, the temporary fire station did not provide ideal accommodation, and he had to work with some old fire appliances such as Green Goddesses and Bedford TKs.

Changes came at Marchwood in August 1988 when the crews moved into a permanent fire station after nearly fifteen years in the temporary accommodation. The opening ceremony was held on 18 September 1989, when the station was opened by Colonel G.F.V. Cowell OBE – Commander Supply South East District. The official programme read:

> The primary role of the Fire Brigade is to provide statutory fire cover when ammunition and other hazardous cargos are being laden or off laden from ships.

In April 1974 a section was established to carry out the role. To meet an increased use of the port a 'B' Type Army Fire Brigade was formed in April 1975. The fire station comprised a Portakabin and a damp open fronted Romney Hut.

The ceremony began at 11.30 a.m. and included the inspection of appliances at the new station. Sub-Officer Derek Turner ran an exercise involving an ammunition fire, and Sub-Officer R. Curtis ran a rescue exercise. There was a buffet lunch to follow. The programme for this event forms a useful historical document as it includes lists of serving personnel, former station officers, those promoted or moved to other fire stations, and a list of guests.

One of the firefighters who serves in the new station is Trevor Fenn. Like Derek Turner, he followed his forebear's footsteps in joining the fire brigade, and this ties Trevor

Newly completed: the first purpose-built fire station at Marchwood Sea Mounting Centre. It has a twin appliance bay, crew accommodation, watch room, and training facilities. (Trevor Fenn)

Colin Fenn joined the fire brigade in Middlesex before moving to ISR Fire Brigade. Here he is driving the Dennis appliance 25 AMX stationed at either Acton or Mill Hill in the early 1960s. (Trevor Fenn)

to local firefighting history. Firstly, Trevor's grandfather was a Blitz fireman in London, later becoming a divisional officer for Middlesex Fire Brigade, and retiring in 1962. Trevor's father, Colin Fenn, also joined Middlesex Fire Brigade. His first day of service was 22 February 1959, and he got married on 24 October that year. Trevor was brought up in Middlesex, but the family eventually moved south, and Colin got a job with Southampton City Fire Brigade at the Docks Fire Station. Whilst there he responded to a job advert for the industrial fire brigade at ISR, and was invited to an interview conducted by Charlie Holmes. Although ISR offered double the salary of SFB, Colin was reluctant to give up front-line service to work in a factory complex, where fire calls would be fewer than in the city. However, Charlie promised Colin a place on the retained crew at Hythe Fire Station within six months if he took the ISR job, meaning that he was happy to accept. It is easy to see how Trevor also became a serving firefighter on the Waterside, and he too served on Hythe's retained crew for a short time.

Trevor Fenn has been a Defence firefighter for a total of twenty-five years. The Army Fire Service became the Defence Fire Service in 1992 with the amalgamation of all armed forces fire brigades, and is now known as the Defence Fire & Rescue Service (DF&RS). Trevor became a fireman and then a leading fireman at Marchwood, transferring to RAF Alconbury – a United States Air Base in Cambridgeshire – for promotion to Sub-Officer. While working there, and also at RAF Mulsworth, he had to wear US military uniform and worked with American fire engines. He transferred back to Marchwood as a Sub-Officer, now known as a Watch Manager, where he is in charge of Blue Watch. Blue Watch is on duty at Marchwood at the same time as White Watch in HF&RS. Trevor has also stood in as station manager for the military base. Marchwood's station manager is overseen by an area manager in charge of '145 Brigade', including the Defence Fire

Medal presentation ceremony held at the Defence Fire Station on 16 June 2009. Trevor Fenn was being awarded the twenty-year Queen's Long Service and Good Conduct Medal, and Alex Petrovic was receiving the Afghanistan Military Campaign Medal after serving there as a firefighter. Left to right: Area Manager Phil McGuinness, Watch Manager Trevor Fenn, Crew Manager Mark Sillett, Chief Fire Officer Noel Roberts, Firefighter Alex Petrovic, -?-, -?-, and Station Officer Russ Symes. (Trevor Fenn)

Stations at Bicester, Middle Wallop, Odiham, and Fleet. This falls under the South Area of the Defence Fire Risk Management Organisation (DFRMO), the others being the North and Central Areas of the UK, and the Overseas Area.

The MOD Station at Marchwood has different risks to cover compared with a local authority fire station. Explosives arrive by rail from MOD Kineton – which also has a fire station – and they are transferred along with vehicles and other supplies to ships normally bound for Germany, Cyprus, Gibraltar, or the Falkland Islands. The main objective with a fire involving explosives is to deploy water as soon as possible, preventing escalation. Originally the Marchwood crews waited on the dockside during loading, a practice favoured by Derek Turner, but they are now asked to remain in the fire station. This is for their personal safety, and also to ensure firefighting personnel would survive an initial blast in order to tackle the aftermath. The dock area has been designed with a 'blast zone' free from buildings, and this is a sign of the danger level involved. The firefighters are always informed when ship loading will occur, but nowadays may carry out training and other duties in the fire station during these times. The station traditionally has run a whole-time watch system, but this was under review in 2013 with proposals to change to day-crewing which was implemented in 2014.

Explosives are not the only risk presented within the dock area at the Sea Mounting Centre, and the training of the fire crews reflects this. They are able to perform rescues from a height, and always stand-by with their vehicle when the cranes are working in case the operator requires assistance. They are also trained in water rescue techniques, operating from a special equipment cabin and the small vessels moored at the dockside. The relevant courses take place at the Manston Training Centre in Kent (the DFRMO has other training establishments at Culdrose, Deepcut, and at Rheindahlen in Germany), and the courses are supplemented by training within the port itself. There is a mock ship's mast so fire crews can practise rescue techniques for incidents on-board ship. Crews can also carry out blue light driver training inside the port complex, once the security gate has been informed by radio. Next to the fire station is a drill tower, training yard, and smoke chamber complex, close to the storage tanks for reserve firefighting water. The DF&RS is unique among the brigades featured in this book in still providing live fire training for its crews. Marchwood's firefighters must cover all the buildings and personnel on the site as well as the materials passing through it.

The MOD fire station has an unusual but fully co-operative relationship with HF&RS.. The two brigades overlap in covering the married quarters, which are outside the perimeter of the main base; the occupants may dial either 999 or an internal telephone number to summon local authority or military assistance respectively! The Marchwood station has a manned watch-room from which to call the county at any time if requiring back-up support. For a fire in Marchwood village, the MOD crew would not be called by Hampshire, but they have in the past responded to nearby incidents, including the Snooker Hall fire in 1991 and a dustcart fire in Normandy Way in 2009. When explosives are not being

The watch room at the temporary build (pre-1988) Defence Fire Station at Marchwood. Staffed continuously, the watch room is always notified when explosives are being loaded at the dockside. (Trevor Fenn)

Firefighters at Marchwood Sea Mounting Centre on a training exercise to practise line rescue techniques. Alan Coles, a temporary Station Officer at Marchwood, is on the stretcher. (Trevor Fenn)

No. 42 KK 14 was a Volvo FL6.14/HCB-Angus appliance issued new to Marchwood Sea Mounting Centre in the early 1990s and liveried for 'Defence Fire Services'. (Derek Turner)

transferred on the dockside, the fire crew is available for first responder duties in the community, and answers these calls for medical assistance using the fire engine! Hampshire and the DF&RS work together for training purposes, and the Southampton fire crews sometimes carry out exercises at the base. There is little involvement between the military fire station and industrial fire brigades in the area, except for the transfer of personnel to take up new positions. For example, Trevor Torrington left Marchwood in 1989 to join the Esso Fire Brigade, and he also served as a retained firefighter for Hampshire, based at Lymington.

Marchwood's firefighting vehicles have altered over the years just as much as local authority vehicles. The earliest livery was a deep red with the royal crest on the cab and the lettering 'Army Fire Service', as carried on old Bedford TK/HCB-Angus appliances. Later, the Defence Fire Service introduced yellow striping for its vehicles. The Carmichael fire engine 67 HF 99 was based at the military port for many years, followed during the 1990s by two successive Volvo FL6.14 appliances with bodywork by HCB-Angus. A Volvo truck complete with the distinctive 'Volumax' cab built by Saxon, VS 68 AA, was based at Marchwood from 2005. It was replaced in March 2011 by the MAN TGM water tender ladder LE 78 AB, the choice of manufacturer reflecting the British Army's move to replace Bedford with MAN vehicles. The firefighters say that the Scottish accent of the in-cab voice announcements reflects the fact the appliance was coachbuilt in Scotland by Emergency One. Its Battenburg livery, with the lettering 'Fire & Rescue' and 'Preventing – Protecting – Responding', does little to distinguish it from the liveries of many county fire brigades. However, it is one of sixty-five pumping appliances currently operated by the DF&RS. Marchwood's second appliance is a Ford Ranger emergency response vehicle, which can be used to carry firefighting foam or to transport emergency equipment for water or rope rescues or hazmat incidents.

The firefighters at Marchwood have been the first line of response to a range of incidents, large and small. An *Echo* article of 9 January 1992 read:

> Army fire crew extinguish blaze at the double: A major alert at Marchwood Military Port this morning turned out to be a minor chimney fire. The alarm, heard in the built-up areas of Southampton and Totton, was sounded at 7.30am but the fire was quickly extinguished by the port's firefighters. Four Hampshire Fire Brigade crews from Hythe, Fawley, Totton and Redbridge also attended.

Without back-up from the county, the MOD crew dealt with an engine fire in a landing craft in July 1992, and a fire in radio equipment in March 1993. They handled an incident involving a chemical lorry, and assisted HF&RS when a person became trapped under a crane in the nearby industrial park. Two of the larger incidents on record took place in 1999. On 22 April, a small heater set fire to a mattress in the Layer Parts Store, and ten Hampshire appliances responded. In August, 100 people, not including local residents, were evacuated when a 1,000lb bomb was dropped from a fork lift during loading. Roads were closed and the ferry operator Red Funnel suspended their operations to the Isle of Wight. Although the bomb did not explode, it sustained a 5mm dent. A major exercise involving multi-agency co-operation took place in September 2011, based on the scenario that

The Ford Ranger emergency response vehicle at Marchwood Sea Mounting Centre is stored empty and can be filled with the appropriate equipment when needed. This kit includes rope rescue equipment, hazmat equipment, and water rescue equipment. (Roger Hansford)

the ship *Hurst Point* had crashed into the jetty, injuring two crew members and leaking phosphorous from its ammunition payload. Personnel from HF&RS and the Defence Fire Station took part in the simulation, known as Operation Spartan. Exercises such as these ensure the area's firefighters remain in constant readiness to deal with a wide range of risks. The risk profile of the Waterside has changed starkly over the last century, but all the local fire brigades continue to show the same dedication as ever.

BIBLIOGRAPHY

Primary Sources

Contract and Plans for the Erection of a Fire Station in Fawley, Shelfmarks H/CL8/525 and H/CA2/3/4 at Hampshire Record Office Archives.

Daily Echo, http://www.dailyecho.co.uk. Accessed 28 September 2012. Records before 2000 are kept in paper form at the newspaper's reference library at Newspaper House, Test Lane, Redbridge, Southampton, SO16 9JX.

Fire Report Books of Southampton Fire Brigade, February 1884–May 1942, SC/F 1/1–4 at Southampton City Council Archives.

HMS Diligence – Fire Cover, Shelfmark H/CL5/FS70A at Hampshire Record Office Archives.

Minutes of Fawley Parish Council Annual Meetings, Shelfmark 25M60/PX1 at Hampshire Record Office Archives.

New Milton Advertiser & Lymington Times, http://www.advertiserandtimes.co.uk/. Accessed 30 September 2012.

Southern Daily Echo.

Waterside Observer.

Waterside Herald.

Secondary Sources

Baker, Eddie, editor. *Buncefield: The Incident and the Operation to Contain It.* Thame: The Fire Brigade Society, 2007.

Brode, Anthony. *The Hampshire Village Book.* Newbury: Countryside Books, 1980.

Browning, Barry. 'Fire Training and Simulators.' *Industrial Fire Journal* (September 2004): 13–17.

Browning, Barry. 'Fire Training & Simulators for Industry.' *Industrial Fire Journal* (October 2005): 5–8.

Browning, Barry. 'Indonesia – Fire & Safety.' *Industrial Fire Journal* (March 1997): 13–15.

Browning, Barry. 'Refinery & Petrochemical Plant Protection.' *Industrial Fire Journal* (March 2004): 19–23.

Browning, Barry. 'Storage Tank Firefighting.' *Industrial Fire Journal* (June 1998): 41–44.

Browning, Barry. 'Testing the Big Guns at Fawley Refinery.' *Industrial Fire Journal* (April 2008): 27–8.

Bunn, Mike, editor. *The Emergency Fire Services in the Second World War.* Thame: The Fire Brigade Society, 2012.

Carter, Colin. 'Rescue Pumps in Hampshire.' *Fire Cover: The Official Journal of the Fire Brigade Society* 195 (May 2012): 11.

Elizabeth II. *Fire and Rescue Services Act 2004: Chapter 21.* London: Stationery Office, 2004.

Fisher, Aidan. *HCB-Angus: Fire Engine Builders.* Stroud: Amberley Publishing, 2012.

Gisby, Brendan and Phil. *The Five Sons of Charlie Gisby: A Family Saga.* Brendan Gisby, 2011.

House, Alan. *Forest Firemen: The New Forest Rural District Council Fire Brigade, 1939-1941.* Southampton: Alan House, 2001.

House, Alan. *Hampshire's Wartime Fire Service, 1939–1945.* Reprint. Southampton: Alan House, 1995.

House, Alan. *Gateway Firefighters: A History of the Southampton Fire Brigade.* Southampton: Alan House, 1996.

House, Alan. *Gateway Fire Engines: The Firefighting Appliances of the Southampton Fire Brigade.* Southampton: Alan House, 1998.

House, Alan and Colin Carter. *Hampshire's Fire Engines: The Rest.* Southampton: Alan House, 2001.

House, Alan. *Home Front Transport: Vehicles of the UK Civil Defence Services, 1938–1968.* Southampton: Alan House, 2009.

House, Alan. *M2HX – 50 Years in Control: A History of Mobilising Fire Engines in Hampshire.* Southampton: Alan House, 1998.

House, Alan. *Proud to Serve: 50 Years of Firefighting and Rescue in Hampshire, 1948–1998.* Southampton: Alan House, 1998.

House, Alan. *The A to F of Hampshire's Fire Engines.* Southampton: Alan House, 2000.

House, Alan. *They Rode Green Fire Engines: The Story of the Auxiliary Fire Service in Hampshire, 1949-1968.* Southampton: Alan House, 2002.

House, Alan. *We Also Build Fire Engines: The Story of Hampshire Fire and Rescue Service Workshops, 1948–1998.* Southampton: Alan House, 1998.

House, Alan. *Wheels of Fire: Fire Engines of the Auxiliary Fire Service and The National Fire Service.* Southampton: Alan House, 2010.

Leete, John. *Under Fire: Britain's Fire Service at War.* Stroud: Sutton Publishing, 2008.

McMonagle, Derek. *Oil & Chemical Processing.* New Edition. London: Esso UK plc., 1993.

Murley, Clare and Fred Murley. *Waterside – A Pictorial Past: Calshot, Fawley, Hythe, Marchwood.* Southampton: Ensign Publications, 1990.

Murley, Clare and Graham Parkes. *Fawley and the Southern Waterside.* Hythe: Waterside Heritage, 2010.

Popham, Hugh. *Esso in Britain: 90 Years of History.* Leicester: Raithby, Lawrence & Co., 1978.

Porter, Tom. *The History of Totton's Firemen.* Tom Porter: Southampton, 1994.

Rolfe, Michael. *Fire Engines.* London: Ian Allan Publishing, 1963.

Sheryn, Hinton. *An Illustrated History of Road Tankers.* Hersham: Ian Allan Publishing, 2001.

Wallington, Neil. *'999': The Accident and Crash Rescue Work of the Fire Service.* Newton Abbot, London & North Pomfret, Vt.: David & Charles, 1987.

Weightman, Gavin. *Rescue: The History of Britain's Emergency Services*. London: Boxtree Limited, 1996.
Whittington, Pam. *Hythe Fire Brigade: A Local History*. Southampton: Itchen Printers Limited, 1998.

Websites

Emergency One Fire Rescue & Emergency Vehicles. http://www.emergencyone.co.uk/. Accessed 5 October 2012.
'Exercise Shannon – Exclusive Preview.' Hemming Fire/Industrial Fire Journal/Firetrade. http://www.hemmingfire.com/news/fullstory.php/aid/1528/Exercise_Shannon_-_exclusive_preview.html. Published 2 May 2012. Accessed 5 November 2012.
'ExxonMobil in the UK.' http://www.exxonmobil.com/UK-English/about_what.aspx. Accessed 19 July 2012.
Hampshire Fire & Rescue Service. http://www.hantsfire.gov.uk/. Accessed 29 September 2012. See also http://www.hantsfire.gov.uk/theservice/specialistresponse/civil-resilience.htm including the dedicated section on HVPs: http://www.hantsfire.gov.uk/highvolumepumping.
Institution of Fire Engineers. http://www.ife.org.uk/about/about/fireengineering. Accessed 13 October 2012.
Kitchen, Iain. *Emergency Service Photos: A Photographic History of the British Emergency Vehicle*. http://www.emergencyservicephotos.co.uk. Accessed 18 June 2013.
Mardon, Roger. 'New Dimension.' http://www.romar.org.uk/page97.html. Accessed 30 September 2012. This website is an excellent resource for Fire Service history.
MOD – Fire Risk Management. www.cfoa.org.uk/download/12875. Accessed 11 November 2012.
Ports and Harbours of the UK: Marchwood. http://www.ports.org.uk/port.asp?id=130. Accessed 8 December 2012.

APPENDICES

Appendix 1

ESSO FAWLEY & MARINE TERMINAL

STATION	APP.	DESTINATION SMALL INCIDENTS 1ST ATTENDANCE	MAJOR INCIDENTS OR MAKE PUMPS 10	MAKE PUMPS 15	MAKE PUMPS 20	OTHER ACTION TO BE TAKEN
FAWLEY	LRP	FIRE	NB(i) For "Major Fires" at Marine	Terminal		1ST ATTENDANCE :- ORDER
FAWLEY	WRL	"	Fire Boat to be mobilised.			FOR FIRE FIGHTING DUTIES.
HYTHE	WRT	"				1 DO Plus 2 other Officers.
BEAULIEU	WRT	"	NB(ii) For Chemical incidents within	ESSO		1 Officer for C U duties.
TOTTON	WRT	"	ET D54 to be mobilised			INFORM :- C.F.O. D.C.O. P.F.C.O.
REDBRIDGE	WRT	"				Duty Officer, D.S.O. S.S. O/c Div.
LYNDHURST	WRT comms	"				Comms Officer. Police. Ambulance.
EASTLEIGH	C U	"				N.5.
FAWLEY	WRT		FIRE			MAJOR INCIDENT/ MAKE PUMPS 10
LYNDHURST	WRL		"			ORDER IN ADDITION TO ABOVE :-
LYMINGTON	WRT		"			1 Officer for fire fighting duties.
REDBRIDGE	WRL		"			1 " " Main Control.
BROCKENHURST	WRT		"			1 " " Forward Control.
LYMINGTON	WRL		ESSO FIRE STN	FIRE		1 " " Safety Officer.
BURLEY	WRT		" "	"		1 " " Foam Duties, to report.
ROMSEY	WRL		" "	"		to Brigade H.Q.
ROMSEY	WRT		" "	"		INFORM :- Tele. Exch. Supervisor.
EASTLEIGH	WRT		" "	"		All A.C.Os.
WOOLSTON	WRT		FAWLEY	ESSO FIRE STN	FIRE	All Divisional H.Qs.
NEW MILTON	WRT		LYMINGTON	" "	"	Police.
RINGWOOD	WRT		BROCKENHURST	" "	"	Ambulance.
WINCHESTER	WRT		LYNDHURST	" "	"	H F B Photographer.
WEST END	WRT		REDBRIDGE	" "	"	H F B Press.
TWYFORD	WRT		ROMSEY		ESSO FIRE STN	MAKE PUMPS 15
NEW MILTON	WRL			FAWLEY	" "	ORDER IN ADDITION TO ABOVE :-
FAREHAM	WRT			LYNDHURST	" "	1 DO Plus 3 other Officers.
CHRISTCHURCH	WRL			LYMINGTON	" "	INFORM :- C.A.O.
COSHAM	WRT			REDBRIDGE	" "	Workshops Eng. Supervisor
RINGWOOD	WRL				LYNDHURST	Ambulance.
HAMBLE	WRT				"	Police.
TITCHFIELD	WRT				"	O.T.B. Brigades.
BISHOPS WALTHAM	WRL				"	MAKE PUMPS 20
GOSPORT	WRT				"	TOTAL OFFICER REQUIREMENT :-
PORTCHESTER	WRT				REDBRIDGE	2 DOs Plus 10 other Officers.
COSHAM	WRL				"	INFORM :- OT B Brigades.
STOCKBRIDGE	WRT				"	Police.
ANDOVER	WRT				"	Ambulance.
BOTLEY	WRT				"	
SUTTON SCOTNEY	WRT				WINCHESTER	Control to contact off duty
ALRESFORD	WRL				"	Control personnel as required.
ALTON	WRT				"	
PETERSFIELD	WRT				"	TELEPHONE CONTACTS
FORDINGBRIDGE	WRT				BEAULIEU	
CHRISTCHURCH	WRT				LYMINGTON	Comms. Centre / ESSO Fire Stn.
POKESDOWN	WRL				NEW MILTON	Fawley 9 892625.
FERNDOWN	WRL				RINGWOOD	Set up R.U.T. with ESSO Fire Stn.
CRANBORNE	WRT				FORDINGBRIDGE	Fawley 9 893656
SALISBURY	WRL				ROMSEY	
WHITCHURCH	WRT				STOCKBRIDGE	Shift Supervisor
HAVANT	WRT				BOTLEY	Day Fawley 9 891558
COPNOR	WRT				COSHAM	Night 9 893183
WATERLOOVILLE	WRT				ALRESFORD	

Chart showing the pre-determined attendance (PDA) at Fawley Refinery & Marine Terminal from the mid-1980s. (Malcolm Rumsey)

Appendix 2

RETAINED MANNING D58 FAWLEY																
DATE	Aug/Sept		6/9		6/9		3/8		1/9		2/9		3/9		4/9	
			MONDAY		TUESDAY		WEDNESDAY		THURSDAY		FRIDAY		SATURDAY		SUNDAY	
NAME	●	*	D	N	D	N	D	N	D	N	D	N	D	N	D	N
BROADHURST	●			✓		✓			✓	✓	✓			✓		✓
SMITH	●		✓		✓		✓		✓		✓	✓		✓	✓	✓
WILLIAMS			✓	✓	✓	✓	✓	✓	✓	✓	✓	✓	✓	✓		✓
WEBB	●		✓	✓	✓	✓	✓	✓	✓	✓	✓	✓	✓	✓	✓	✓
PARRATT	●	*	✓	✓	✓	✓	✓	✓		✓		✓	✓			✓
RUMSEY	●	*		✓		✓		✓		✓		✓	✓	✓	✓	✓
RUSH	●					✓	✓	✓		✓	✓	✓		✓		✓
WYATT	●			L	E		A		V		E					
DEAN	●			✓		✓	✓	✓						✓	✓	✓
RIXON	●	*	✓		✓			✓	✓	✓	✓	✓	✓	MKG	✓	
HORNE	●			✓		✓		✓		✓		✓				
TOTALS			5	8	5	8	6	7	5	9	6	9	5	7	6	8

● ....B/A
* ....RED DRIVER

SIGNED BY ....

Chart showing planning for retained cover at Fawley, to ensure there was always a crew, with BA wearers, and a driver. (Malcolm Rumsey)

Appendix 3

HAMPSHIRE FIRE BRIGADE

Have you ever thought of yourself as a part-time member of the Hampshire Fire Brigade?

The Hampshire Fire Brigade is a public service provided to protect the people of Hampshire against fire and other emergencies. The firefighter tackling this stimulating and challenging job is justly proud of his service and of the status which his team enjoy in their local community.

The importance of Part-time Firefighters

It would not be practical to provide fully employed wholetime firefighters at all of Hampshire's 55 fire stations. In fact 38 of those stations in the lesser populated areas of Hampshire are manned solely by part-time firemen who number over 700 and who perform a valuable function.

In 1976 they attended and dealt with over 11,000 incidents of all types including serious fires, road traffic accidents, flooding and many other humanitarian services. When summoned to his fire station the firefighter will never know which of his many and varied skills is about to be employed.

What is expected of the Firefighter?

Part-time firefighters work at other employment and attend fires when called. To comply with this arrangement they must either live or work and preferably both, within ½ of a mile of their local fire station. The job is particularly well suited to shift workers or those with employers sympathetic to the Fire Brigades needs and who are willing to let their employees help to protect their community. Availability during most working days is essential for the system to succeed but the Fire Brigade is very understanding about family and job commitments. Guaranteed attendance to every call is not required, but a reasonable response is of course expected.

New entrants to the part-time service should be between the ages of 18 and 40 (possibly older if fully fit), at least 5 feet 6 inches tall, have good physical fitness and eyesight and be of smart bearing.

Training

New members are provided with a uniform and full personal firefighting kit. They undergo a paid course on three successive Sundays at the Brigade Training Centre. On satisfactory completion of the course they are provided with a modern pocket alerter which allows freedom of movement over a wide area.

Promotional material for becoming a retained fireman in the 1980s. (Malcolm Rumsey)

INDEX

Personnel with greater significance in the text have their own index entry; others appear under the fire brigade in which they served. They are listed with their most senior rank mentioned in the book.

If you enjoyed this book, you may also be interested in…

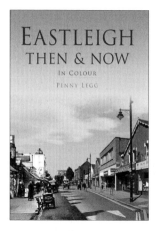

Eastleigh Then & Now
PENNY LEGG

The old Saxon village of 'East Leah' was recorded as existing as early as 932, situated on the ancient Roman road between Winchester and Bitterne. By the mid-nineteenth century, the town was called Eastly, but, with the coming of the railway it was renamed Eastleigh. This stunning full-colour volume uses archive images, contrasted with modern-day photographs, to show how Eastleigh has changed over the last century. It seeks to capture the present–day town, before further change comes to the area and reshapes it forever.

978 0 7524 6999 7

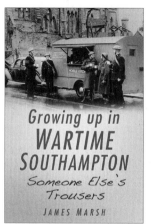

Gowing Up in Wartime Southampton
JAMES MARSH

This is a story spanning some of the most turbulent decades in recent world history. James Marsh was born during the first year of the Second World War, and many of his infant years were spent in air-raid shelters. Bombs rained down from the German Luftwaffe as they tried to destroy the city of Southampton. The gritty determination, community spirit, and above all, the humour, with which the local community faced the difficulties of war have stayed with James throughout his life.

978 0 7524 5658 4

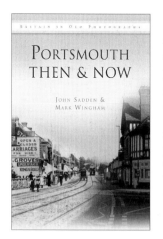

Portsmouth Then & Now
JOHN SADDEN

The major port and popular tourist city of Portsmouth has a rich heritage which is uniquely reflected in this fascinating new compilation. The photographs in this absorbing collection enable the reader to explore the differences that passing time has wrought on the streets, neighbourhoods, businesses, houses and, not least, the people of Portsmouth. Inspiring fond memories in some and revealing the Portsmouth of yesteryear to others, this volume will appeal to all who know this ever-changing city.

978 0 7524 5658 4

Visit our website and discover thousands of other History Press books.

www.thehistorypress.co.uk